THREE BIG BANGS

Three Big Bangs

Matter-Energy, Life, Mind

Holmes Rolston III

COLUMBIA UNIVERSITY PRESS NEW YORK

COLUMBIA UNIVERSITY PRESS

Publishers Since 1893

New York Chichester, West Sussex

Library of Congress Cataloging-in-Publication Data

Rolston, Holmes, 1932–

Three big bangs : matter-energy, life, mind / Holmes Rolston III.

p. cm.

Includes bibliographical references (p.) and index.

ISBN 978-0-231-15639-4 (cloth : alk. paper) — ISBN 978-0-231-52684-5 (electronic)

1. Evolution. I. Title.

B818.R65 2011

113—dc22 2010003254

Columbia University Press books are printed on permanent and durable acid-free paper.

This book was printed on paper with recycled content.

Printed in the United States of America

c 10 9 8 7 6 5 4 3 2 1

CONTENTS

ILLUSTRATIONS

PREFACE

Plato argued, famously, that we ought to "carve nature at the joints" (*Phaedrus* 265e). I plan to "carve nature at the explosions." There have been three big bangs: generating matter-energy, generating life, generating the human mind. These explosions form no simple continuum but a complicated, diffracted, exponential story. "Big bang" is here a metaphor for critical, exponential, nonlinear bursts with radical consequences for exploring new state spaces with novel combinatorial possibilities. Using another term, there have been three "big singularities."

En route we will have to use our big heads to avoid what I might call the rarification fallacy, the idea that what is rare is unimportant. Are the singularities signals revealing the nature of nature? Transcendent Presence in, with, and under nature? The three big bangs raise big questions, ultimate questions.

Seeking "theories of everything" is tolerated, even encouraged, within astrophysical cosmology, but if a metaphysician seeks a "grand narrative," that is often discouraged as an impossible quest. The world is too big and too complex even to think about all in one go. The best we can do is philosophy piecemeal. Equally, anyone who seeks "foundations" is nowadays thought to be hopelessly archaic. My search here is "archaic" in a better sense, a search for the elemental, fundamental givens we encounter in nature. Finding these, what worldview can we build?

Socrates claimed: "The unexamined life is not worth living" (*Apology* 38). "Know thyself." Yes, but: "Life in an unexamined world is not worthy living either." Humans are the only species capable of realizing how grand is this world they inhabit, the astronomical universe, the panorama of life that vitalizes this planet, as well as of examining themselves and the minds with which they do so. One needs encounter with the nature of these three big bangs—matter-energy, life, mind—to become a three-dimensional person. Foundations or not, one does need to be inclusive, comprehensive in one's worldview. Today, the examined life is not possible without examining the world in terms of these three big bangs.

Discovering the three big bangs has required the genius of millennia of human history; today we see further because we "stand on the shoulders of giants" (Newton). At the same time, the nature we have discovered with such collective genius is inescapably right there "in your face" in the great outdoors for anyone to see. A walk in a forest confronts you with life surrounding; the starry night sky is over your head. Facing life surrounding, facing the night sky, you realize mind behind your face: the facing mind.

We can take Albert Einstein as an icon of discovering the first big bang in the astronomical heavens (or at least of contemporary physics); we can take Charles Darwin as an icon of discovering the second big bang, evolutionary life on Earth. But then the third big bang inescapably confronts us. Continuing to take Einstein and Darwin as icons, the marvel is not just heavens above or earth beneath; the marvel is equally, indeed more so, the human minds capable of such knowledge.

This is a short book for searchers, students, and academics, and also for the general literate reader. In this search, I make full effort to put readers in contact with the original sources and authorities. We here try to look over their shoulders to see what those in the know know. Or, since at cutting edges the experts often still wonder, where they think the intellectual frontiers are. We humans today, in a new millennium, seem poised on the edge of yet another combinatorial information explosion, with escalating possibilities in science and technology, evidenced in the recent decoding (and possible transforming) of our own genome and in unprecedented information storing, sharing, and processing on the Internet. Wondering where we are, wondering who we are, we will be better able to wonder what we ought to do.

No one can know firsthand all the details of the materials we survey; that would require a mastermind in cosmology, microphysics, evolutionary history, paleontology, genetics, molecular biology, neuroscience, psychology, logic, ethics. Still, I write with the persuasion that good philosophers and theologians, good inquiring minds, can look over the shoulders of those who are doing these things and spot what is metaphysically interesting. In one sense, doing this, we are jacks of all trades, masters of none. In another sense, we are more inclusive, more comprehensive than the scientists.

Then again, whatever the philosophers may say with their epistemological doubts, the sciences do demand more accounting of these singular data points: matter, life, mind. The scientists have developed strategies—theories, instruments, critical methods—that do approximate a real world. Astrophysics has discovered a vast universe, deep space and time; evolutionary biology has documented a vast fossil record. Whatever one thinks about the social construction of science, it is irrational to deny that such discoveries are of what is objectively there: galaxies and fossils; and that these record a long and eventful past that precedes us and makes our contemporary cognizing presence possible.

Both the first and the second big bangs resulted in us: the *Homo* that is so *sapiens*. A third has taken place within us, the mind's big bang in the explosion of cultures with radical capacities for the generation and cumulative transmission of ideas—knowledge, wisdom. "Man is the measure of things," said Protagoras, another Greek philosopher (recalled in Plato, *Theaetetus*, 152). Perhaps better: "Humans are the measurers of things." We here seek to take the measure of heavens above and earth beneath. That will force taking measure of ourselves.

The measure of all three is radical genesis; in the end we must ask the questions with which Genesis begins: wondering about creation resulting in persons who image God. We are better placed than any generation in human history to ask these questions, to take the measure of them. But answers to ultimate questions still lie almost beyond our reach. If found, answers will focus on the three big bangs.

Boom! Boom! Boom! Be forewarned: the territory ahead is intellectually explosive.

CHAPTER 1

The Primordial Big Bang

Matter-Energy

Scientists have been discovering astronomical deep space and deep time, as well as pushing "deep down" from molecular to micro nature. Time has expanded to almost forever, the universe expanded to staggering, inconceivable distances across intergalactic space. We now know of phenomena at structural levels from quarks to quasars. We measure distances from picometers to the extent of the visible universe in light-years. We measure time from picoseconds to the billions-of-years age of the universe. Putting such discoveries together, we have found dramatic interrelationships between astronomical and atomic scales that give us a startling picture of our physical universe. At some risk of vertigo, let's explore the explosive startup and ongoing expansion of this lavish universe—the first big bang.

Do the macrophysics and the microphysics affect our metaphysics?

Explosive Startup and Ongoing Expansion

The universe is expanding through the stretching of space between galaxies, and if one runs the history rearward, it shrinks to a point of startup, a primordial hot big bang, now dated about 13.7 billion years ago. Cosmologists have made claims about the duration of the "early universe"—from about one microsecond after the initial "singularity" (as they may call it) to several hundred thousand years (400,000 years) as a superhot universe. They make further claims about an initial "inflation" when the originating universe, less than 10^{-35} seconds old, jumped in size by an enormous factor (about 10^{60}), expanding faster than the speed of light (Guth 1997; Linde 1990). At that time all of the present universe was somehow packed into space smaller than that of a typical atom, and this at extremely high temperature.

Cosmologists can wonder if their capacities to describe what was going on under these conditions are credible, since most of the astrophysical and microphysical processes we otherwise know would there break down. Most rather doubt that we can know anything before what they call Planck time, when the universe was 10^{-43} seconds old (though see below on symmetry-breaking). Perhaps even time itself appears somehow in the startup out of an initially atemporal big bang, so that to think of some "time zero" is misleading. Nevertheless, scientists are at near consensus about an originating singular huge explosion.

This earliest point of the present universe was tagged "the big bang" by Fred Hoyle, first derisively, but the name stuck. So the first big bang is an explosion of matter-energy. The creation of something out of nothing at the beginning of time was clearly a remarkable occurrence. Or if there was something before, or if the "nothing" was some sort of creative vacuum, the explosion was still spectacular. There was explosion, inflation compounding explosion, and continuing explosion after that. That is superexplosion. The result is huge: if in some spaceship we could travel at the speed of light, it would take us billions and billions of years to cross it. This huge universe resulted from an explosion starting as a tiny speck. (Actually all this was silent. "If a tree falls in the forest, and there is no one to hear it. . . .")

Since the big bang, whether explosion continues is a matter of perspective. The universe is still expanding, though at a diminished rate. There is some evidence that the presently continuing expansion is speeding up. There was creativity at the primordial big bang, which launched ongoing creativity in the expansion. Here the expansion rate (the ongoing explosion rate, if you like) proves critical. If the expansion rate of the universe had been a little faster or slower, then the universe would already have recollapsed or the galaxies and stars would not have formed. This expansion rate figures into astronomical calculations such as those for the strengths of the four fundamental forces, or of the cosmological constant. We return to the often puzzling character of these astronomical facts below.

From one perspective, in the big bang, everything is flying apart in a universe continually expanding and generally uniform (isomorphic); but from another perspective, there are local departures from the overall smoothness. In these non-isomorphic regions, under the influence of gravity, matter clumps up into stars, into galaxies, the loci of ongoing creativity. The particulars of such stars and galaxies may depend on earlier random fluctuations, perhaps even quantum indeterminacies. Or they may depend on the intersections of previously unrelated causal lines (stars crashing into each other) or involve chaotic features. But the overall processes are nomothetic, lawlike (making celestial mechanics possible, or explaining stellar evolution).

This energetic matter not only clumps, it complexifies. Nature aggregates and builds. Across this long time span in the rapidly expanding universe, the stars are the furnaces in which all but the very lightest elements are forged, a process called nucleosynthesis (Clayton 1983). Further, the various heavier elements—carbon, oxygen, sulphur, nitrogen, silicon, all of the elements heavier than hydrogen and helium (also when the universe was still hot, some lithium)—are synthesized in proportions that make later planets and life possible. These elements, made of protons, neutrons, electrons, inner positive nuclei and outer negative shells, are forged with bonding capacities, almost like grappling hooks, making possible endless recombinations. The stars run their courses and some explode as supernovae, dispersing the heavier elements from their production sites throughout space. Such matter is condensed as planets, and life evolves out of such elements.

From our present human perspective, the cosmic big bang may still seem like a lot of waste—all those galaxies, stars, asteroids, cosmic dust, dark matter, dark energy. Do we really need a universe with a hundred billion galaxies, each with a hundred billion stars? Maybe we are lost out there in the stars. Explosions make a lot of noise (loud sound from the burst of power), a lot of "noise" (chaotic background disturbance confusing any signal). Well, maybe the first big bang did not make a sound, but is there any signal in the scattering of galaxies, stars, asteroids, black holes, and so on? Even if much of the bang was meaningless noise, from this "singular" universe, some of the results of the huge explosion continue as the Earth, the earth, the dirt under our feet, the flesh and blood of our bodies and brains. And we have no scientific theory as to how we might have obtained such bodies and brains without some remarkable elemental source, such as this singular big bang provides.

The universe is so huge that we can see only the parts of it in our light cone, the area within which the light has had time to get to us. The Hubble Space Telescope has imaged galaxies over 10 billion light-years distant. But if the scale of the universe were much reduced (to galaxy size for instance, 100,000 light-years across), there would not have been enough time for stars to form and generate the elements beyond hydrogen and helium, elements that later make life possible. John Barrow surveys the universe: "Many of its most striking features—its vast size and huge age, the loneliness and darkness of space—are all necessary conditions for there to be intelligent observers like ourselves" (Barrow 2002:113). There is, in fact, a lot that is quite singular resulting from this singularly huge explosion.

A Singular Universe: Unfolding Order

Explosions can be rather messy, often more disruptive than creative. Our exploding universe, however, settled into an expansion rapid enough to prevent collapse on itself and slow enough to continue for billions of years, permitting galaxies, stars, and planets to form. So, despite this explosive character of the infant universe, the result produces much order. The astronomical (and terrestrial) realms of matter-energy are lawlike—mechanistic, a clockwork universe. The initial conditions at the big bang were, presumably, idiographic, that is, unique to this universe. But those fermenting conditions produced a coherent universe that is physically nomothetic, lawlike, and ordered, whatever the chaotic elements.

Scientists typically claim that the laws of physics and chemistry are true and unchanging all over the universe, often in contrast with the biological sciences, where it may be claimed there are no universal laws at all, only generalizations in an earthbound natural history. Physicists may indeed brag that theirs is the most ordered of the sciences, with more mathematical and logical rigor than biology, psychology, sociology. Physicists can predict eclipses centuries hence; economists cannot predict the stock market tomorrow. Think of all the equations in physics, such as $E=mc^2$, or the reactions in chemistry, based on the atomic table posted behind the lecture podium in every chemistry classroom. The impressive rigor of physics and chemistry is seen in their metric character, with accompanying predictability and testability.

All the mathematics underlying the discussions to which we soon turn regarding the "theory of everything" or the "fine-tuned" character of the pivotal processes in the formation of the stars, elements, and planets underscores this order. Einstein concluded, famously, that "the eternal mystery of the world is its comprehensibility" (Einstein 1970:61). Eugene P. Wigner, a physicist and mathematician, contends "that the enormous usefulness of mathematics in the natural sciences is something bordering on the mysterious and that there is no rational explanation for it. . . . The miracle of the appropriateness of the language of mathematics for the formulation of the laws of physics is a wonderful gift which we neither understand nor deserve" (Wigner 1960:2, 14).

John A. Wheeler exclaims, "This is a world of pure mathematics and when we penetrate to the bottom of it, that's all it will be" (Wheeler, interviewed in Helitzer 1973:27). Is there nothing but order, captured by mathematical precision, a "matheomorphic" universe, as though the big bang was actually a mathematical explosion? Something is needed beyond the pure mathematics to compel it to exist in an actual world. There are worlds imaginable in pure mathematics that are never realized. Though Stephen Hawking delights in searching for a theory of everything, he goes on to ask:

> Even if there is only one possible unified theory, it is just a set of rules and equations. What is it that breathes fire into the equations and makes a universe for them to describe? The usual approach of science of constructing a mathematical model cannot answer the question of why there should be a universe for the model to describe. (Hawking 1998:190)

The theoretical, matheomorphic universe exploded in an actual fireball, and the fire still burns after thirteen billion years.

The mathematical character of high-level physics, even after we can no longer picture what is going on, does suggest that the ordered intelligibility of this exploding, expanding universe vastly outruns our sensory capacities for perception and our local capacities for experience. The mathematics still seems to contact and correspond to physical nature. As far as our capacities for thought reach, whether in words or in mathematics, the universe seems unreasonably "reasonable," intelligible, despite the fact that we can no longer visually represent, verbally model, or perceptively sense it. The math still works even in realms where sense and intuition do not easily serve. The explosive big bang produces a realm of exquisite, supersensory rationality that transcends but supports sense, space, and time. We are not yet prepared to consider the third big bang: mind. But perhaps there is already an intimation. Mathematics is, above all, mental; it is the logical creation of the human mind, and the fact that mathematics repeatedly helps us to understand the structure of the physical world corroborates the belief that the world we inhabit is the creation of mind. We might even need that encouragement when we plunge into the chaos of biology.

The big bang launched natural history. In the ongoing explosion, mathematics remains powerless to appreciate a world until it adds a narrative of events. Perhaps in advanced physics, there are only equations, with no pictures, but mathematics is useless without a text, without words—no matter how much it is also true that mathematics accomplishes what words cannot. The spacetime diagrams must have a caption, the equations an interpretation. Past this, complex nature is never fully described by mathematical models. To the contrary, very much is left out, and mathematics is to that extent stylized and crude as a description of rich natural processes. Its precision is bought with its incompleteness. Neither mathematics nor other forms of physics anywhere know the categories of life and death, nor mind and conscious experience, which, with the second and third big bangs, became the phenomena that most cry out to be explained. Even within physical cosmology, there are factual claims such as those involving the anthropic principle—observations about values of fundamental constants, forces, conditions that are prerequisites for the complex chemistries of life. These may be mathematical, based on values in equations, but the cosmological interpretation of these facts is not. The interpretation is historical, metaphysical, theological.

Nature has mathematical dimensions at every structural level. But we do not from this conclude that all the world's cleverness and beauty lie in its mathematics. Even if we were to lay aside the upper levels that metricize less well, at the quantum levels our metricizing capacities, profound as they are, run to an end zone. We cannot completely metricize the individual quantum event; it defies mathematical specification in its concreteness. At this point, curiously, one of the most impressive of our mathematical theories tells us that nature permits no further mathematical specifiability.

Certainly the order is impressive, nowhere more so than in the mathematics that maps the big bang. But mathematics is not the only mode of thought competent for judging multidimensional nature. Physics and chemistry are the most abstractive of the sciences. To some extent they are abstracted out of a more messy real world: physical laws are not so much ultimate and absolute as they are approximations over statistical averages with margins of error. Impressive as such laws are in physics and chemistry, they leave out all the emergent eventfulness with which the other sciences and the humanities will want to deal. Physics and chemistry take no special subset of natural entities for their subject matter, while biology takes organisms, psychology takes behavior and mind, sociology takes societies, and even the special physical sciences—geomorphology or meteorology—have their restrictions. We need to stay alert to the paradox that these universal physical sciences, which seem so powerful in interpreting what has resulted from the primordial explosion, also drastically oversimplify (Ellis 2005).

There is yet another side to this emphasis on order unfolding from the primordial big bang, especially when we anticipate what kind and levels of order make possible the second and third big bangs. We do not find physically, nor do we want philosophically, any law that says: order, more order, more and more order. Logically and empirically, beyond mathematical order and predictability, there must be an interplay of order and disorder, certainty and openness if there is to be autonomy, freedom, adventure, success, achievement, emergents, surprise, and idiographic particularity.

Order is related to information, and we will in the next chapter be analyzing this in biology, where it is a central theme in genetics, after the second big bang. Today, with the exploding that has resulted in the third big bang and our ever-advancing human cognitive capacities, we may think we have entered the information age. Information theory began in electronics and computing, and physicists sometimes ask about the information content of the physical world. Hans Christian von Baeyer, a physicist, anticipates: "If we can understand the nature of information, and incorporate it into our model of the physical world . . . then physics will truly enter the information age" (von Baeyer 2003:17).

John Wheeler, following from his claims that the world is pure mathematics, has made a further famous claim, enigmatically epitomized in his aphorism "it from bit." The world of objects, "its," is rooted fundamentally in "bits," information units, a term borrowed from computer memories. "It from bit symbolizes the idea that every item of the physical world has at bottom—at a very deep bottom, in most instances—an immaterial source and explanation . . . in short that all things physical are information-theoretic in origin and this is a *participatory universe*" (Wheeler 1994:296). "'Getting its from bits' . . . refers to a vision of a world derived from pure logic and mathematics" (Wilczek 1999:303).

Wheeler speculates that order penetrates the universe as a sort of network or circuit loop, even involving backward causation, in which there is a Platonic demand for intelligibility. The physical world gives rise to the possibilities of communication; intelligent agents evolve, who analyze nature and find it rational, mathematical. But these agents, though coming later in time, are determinants of the physical characteristics of the universe. "The whole show is wired up together." "Will we someday understand time and space and all the other features that distinguish physics—and existence itself—as . . . a self-synthesized information system?" (Wheeler 1999:316, 321). At this point, however, Wheeler goes beyond his "pure mathematics." He does need, in his metaphor, to get existing "its" (actual physical objects) from his "bits" (mathematical forms). The "its" of the real world are interparticipatory with "bits" of significance, meaning.

In another metaphor, continuing the idea of a self-synthesized information system, the universe is sometimes described as a computer. In various cultures, nature has been described with diverse metaphors: the creation of God, the Great Chain of Being, a clockwork machine, chaos, an evolutionary ecosystem, Mother Nature, Gaia, a cosmic egg, *maya* (appearance, illusion) spun over *Brahman*, or *samsara* (a flow, a turning) which is also *sunyata*, the great Emptiness, or *yang* and *yin* ever recomposing the *Tao*. Our culture lives in the computer age, and so perhaps "computer" is just the latest in such shifting models.

Still, the model might give some insight. The "computational universe" is programmed, as it were, to start simple and generate complexity, in the course of which it generates intelligent output, including life and mind (Lloyd 2006). Some scientists are impressed with the capacities of simple systems with a few basic rules (algorithms) to generate complex patterns (Wolfram 2002; Weinberg 2002). Such a process is more cybernetic (in the language of information theory) than it is mechanical (in classical Newtonian language). If the universe is a machine, it is still more fundamentally an information-processing system, a system tending toward generating information. The physical world at first appears to be nothing but causation, A causes B causes C, the mechanical world. But more is going on. That becomes so obvious in the second and third big bangs, biology and mind, that we need to detect it already ticking in the first big bang, physics. Molecules in space are mostly inorganic, for instance, but some of them are already prebiotic—such as carbon molecules and even amino acids (especially in the 1969 Murchison meteorite; Kvenvolden et al. 1970). That suggests that mechanical atoms have some possibility of self-assembly into biological molecules.

But the computer metaphor may equally have its limits, parallel to those we worried about when thinking of the world as pure mathematics. The limits are in two directions. The analogy may be too strong. The physical universe, so far as it is lawlike and predictable, may be only a realm of causation, not one of computing. Or, astronomical natural history may have locations that are too random, too indeterminate, too piecemeal for the whole to form a computer.

The analogy may be too weak. Even more revealingly, comprehensive natural history may be too emergent, too surprising, too narrative to be computable. Perhaps one can compute from big bang to galaxies to stars to planets. But can one compute across singularities, especially across the three big bangs? One cannot compute from trilobites to elephants to self-conscious humans, even if the DNA sequences that make this evolution possible are digital. There is too much story, history; computers are not good at generating or detecting plots. "Information" is a richer category than "computing."

The term "information" is complex and has been used variously in differing sciences. There is information on the surface of the moon, in the sense that a geologist can read some of the history of the moon from the overlay of meteoric impacts there. There is information in the cosmic background radiation, in the sense that cosmologists can backtrack from radiation data and draw conclusions about the early history of the universe. Mathematical information in communication theory, Shannon information, deals with reliable signal transmission, without regard to the significance, the semantic content, of the signal transmitted. Relevant information in addition has both signal reliability and signal significance. Any science, physics included, is a question of information gained.

But "information" in such use does not refer to any objective knowledge in physical systems, absent human scientists, nor to any analogue or predecessor of "information" of the kind that does evolve later in biological systems. In genetic coding in DNA, the significance or semantic content of such information is critical, as it was not in the minimal, mathematical, physical sense of information. Hubert P. Yokey insists: "Life is guided by information and inorganic processes are not." "The sequence hypothesis in the genome and in the proteome is a new axiom in molecular biology . . . unique to biology for there is no trace of a sequence determining the structure of a chemical or of a code between such sequences in the physical and chemical world" (Yokey 2005:8, 183). If we find no trace in the physical and chemical world of that which is the dominant form of order in the biological world, we will be forced to think of the second big bang as a serendipitous singularity—except insofar as the phenomena described in the anthropic principle lead us to wonder about a readiness for life already present in the physical and chemical materials. We are left puzzling about the extent and origins of order.

Logical Explosion: A Theory of Everything?

Cosmologists hope to arrive at what, with a mixture of hope and jest, they have since the mid-1980s called "a Theory of Everything" (Barrow 2007; Tegmark 1998). R. B. Laughlin and David Pines explain: "The Theory of Everything is a term for the ultimate theory of the universe—a set of equations capable of describing all phenomena that have been observed, or that ever will be observed" (Laughlin and Pines 2000). Physicists speak of a Grand Unified Theory (GUT) that would unify in a single model the various theories of the fundamental interactions and laws in physical nature. This continues their conviction, following Einstein and Wheeler, that the universe is mathematically (if also mysteriously) reasonable, that with mathematics scientists might get to the bottom of things.

No such theory exists today; but if such a world formula comes, it might explain the explosive big bang. A final theory of this scope would specify our particular universe, accounting for its fundamental characteristics in a detailed and inclusive way. Pierre Simon, Marquis de Laplace, a French mathematician and astronomer, once famously claimed that a sufficiently powerful intellect, well positioned in the early universe, knowing the laws of nature and the positions and velocities of all particles, would know the future, predicting it from the past.

There are two fundamental theories of physics: quantum field theory and general relativity. Quantum field theory takes quantum mechanics and special relativity into account, a theory of all the particles and forces, but it ignores gravity, which is, of course, a principal force in holding the world together. General relativity is a theory of gravity, but it ignores quantum mechanics. No one knows at present how to reconcile the two; but, if discovered, a theory of everything might do it. The most popular version currently is some form of superstring theory. If physicists find such a theory, then we might be able to claim that the universe is such that it must have produced not only these remarkable results of the first big bang, but equally those of the second and third big bangs—similarly to what is sometimes called a strong anthropic principle.

However, despite our search for some theory of everything, the big bang itself resulted from no known laws in physics; it too is a singularity. Even if there are multiple universes (see below), there is nothing in physics that predicts this particular exploding universe with its laws, constants, instant inflation, initial conditions, and particular arrow of time. "The simple and absolutely undeniable fact is that the universe did not have to have the particular laws it does have by any logical or mathematical necessity" (Barr 2003:148).

A first consideration is to keep any such claims about a theory of everything under some logical and empirical control. Four fundamental forces hold the world together: the strong nuclear force, the weak nuclear force, electromagnetism, and gravitation. What is mainly sought is a theory that would unify these four fundamental forces and would also explain the existence and transformations of different kinds of particles, perhaps also the values of the fundamental constants. Nothing in such a nomothetic (lawlike) theory would explain the idiographic (particular) details of the actual world in its ongoing dynamism over the last thirteen billion years—why in our solar system Saturn has rings, Jupiter a great Red Spot, and Earth an ocean and tectonic plates. A unifying theory of elemental fundamentals would be too low-level, too basic to explain anything at all about the behavior of complex systems, such as genetic coding in organisms, species in ecosystems, economic forces in capitalist society, or voting patterns in a national election. So Laughlin and Pines, after giving the definition above, back off: "The Theory of Everything is not even remotely a theory of everything," since it cannot begin to deal with biologically emergent behaviors and complex adaptive systems (Laughlin and Pines 2000). Any physical theory of everything is a thousand orders of magnitude away from a philosophical theory of everything. Any theory of everything only explains fundamental processes in the primordial big bang but does not begin to reach the second or third big bangs.

Philosophically, even if physics did produce a "theory of everything" that made developments from the big bang to human culture inevitable (singularities included), that would be even more remarkable and further support the argument that mind is built in, with, and under the process.

In a pivotal early paper investigating the anthropic principle, Bernard J. Carr and Martin J. Rees concluded:

> The possibility of life as we know it evolving in the universe depends on the value of a few basic physical phenomena—and is in some respects remarkably sensitive to their numerical values. . . . One day, we may have a more physical explanation for some of the relationships discussed here that now seem genuine coincidences. . . . However, even if all apparently anthropic coincidences could be explained in this way, it would still be remarkable that the relationships dictated by physical theory happened also to be those propitious for life. (Carr and Rees 1979:612)

Martin Rees continues, two decades later:

> Perhaps a fundamental set of equations, which may some day be written on T-shirts, fixes all key properties of our universe uniquely. It would then just be an unassailable fact that these equations permitted the immensely complex evolution that led to our emergence. But I think there would still be something to wonder about. It is not guaranteed that simple equations permit complex consequences. (Rees 2001:162)

Stephen M. Barr puts the point this way, with emphasis:

> Even if all the physical relationships needed for life to evolve were explained as arising from some fundamental physical theory, *there*

> *would still be a coincidence*. There would be the coincidence between what that physical theory required and what the evolution of life required. If life requires dozens of delicate relationships to be satisfied, and a certain physical theory also requires dozens of delicate relationships to be satisfied, *and they turn out to be the very same relationships*, that would be a fantastic coincidence. Or, rather, a series of fantastic coincidences. (Barr 2003:145)

A Biogenic/Anthropic Universe

Paul Davies, a cosmologist, claims that we hit "the cosmic jackpot," a universe "just right for life" (Davies 2007). Phrased more technically, "virtually no physical parameters can be changed by large amounts without causing radical qualitative changes to the physical world. In other words, the 'island' in parameter space that supports human life appears to be quite small" (Tegmark 1998:6). The range of values of some quantity that is life-permitting (such as the strength of the force of gravity) is small compared with the range of values that physical theory might otherwise allow. Although physics finds such a jackpot universe just right for life, theories in physics, as we were noticing, nowhere require that there be such a universe. The question now is not so much whether cosmologists can explain "everything" as whether they can explain how there comes to be "anything"—especially anything as interesting as (in Davies' metaphor) "a cosmic jackpot."

In the last half-century cosmologists have found dramatic interrelationships between astronomical and atomic scales that connect to make the universe "user-friendly." These discoveries are commonly gathered under the name "the anthropic principle," a term introduced by Brandon Carter in 1974, though it could better have been named "the biogenic principle." Nor is this even a "principle" in any familiar sense. Rather the reference is to a series of observations about the values of the fundamental constants, the fundamental forces, the properties of particles, such as charge and mass, the nature of dynamic processes, the initial conditions. When these are figured into the theories of physics, the result in our universe appears to be "fine-tuned" so as to enable the development of complex chemistries, which are requisite for life (the second big bang) and self-conscious mind (the third big bang).

How the various physical processes are "fine-tuned to such stunning accuracy is surely one of the great mysteries of cosmology," concludes Paul Davies. "Had this exceedingly delicate tuning of values been even slightly upset, the subsequent structure of the universe would have been totally different." "Extraordinary physical coincidences and apparently accidental cooperation . . . offer compelling evidence that something is 'going on.' . . . A hidden principle seems to be at work" (Davies 1982:90, 110).

Physicists cannot do experiments revising the universe, but they have been doing thought experiments to see whether another one would be more congenial. Such "if-then" experiments conclude that the universe is mysteriously right for producing life and mind. We next turn to some half dozen of these considerations, among fifty or sixty that have been variously explored (Barrow et al. 2008; Davies 2007; Rees 2000, 2001; Barr 2003; Denton 1998; Barrow and Tipler 1986; Leslie 1989).

(1) The rate of expansion of space in the universe depends on the cosmological constant, usually symbolized by the Greek letter lambda (λ), which is quite small (nearly zero but not zero) (Vilenkin 2006). The cosmological constant is a term that Albert Einstein introduced into his theory of general relativity when applied to the universe, thinking thereby to allow the universe to remain static (nonexpanding, nonshrinking). Lambda is related to the acceleration of the universe. If this constant is positive, the universe expands. If negative, the universe contracts. Over a decade later Edwin Hubble discovered that, despite any effects of gravity, the universe was in fact expanding, based on observations of redshift in starlight from distant galaxies (the Doppler effect).

Interestingly, the expansion rate of our "user-friendly" universe depends rather precisely on this minute constant. Expressed in natural units, it is less than 10^{-120}. Written as an ordinary decimal, this would be:

0.00000000000000000000000000000000000000
0000000000000000000000000000000000000000
0000000000000000000000000000000000000001

The small but positive cosmological constant can be thought of as a pressure, associated with an energy in empty space, that overcomes gravitational tendencies toward contraction and results in accelerating expansion. "Empty space" in cosmologists' terms is empty of ordinary matter; they may call it a vacuum, but it is not really nothing. It is a sort of energy possibility space. Empty space is supposed to have latent particles that can pop into and out of existence, perhaps to contain dark energy and have other activities. Cosmologists speak of a space-time "quantum foam" or of microscopic quantum fluctuations in empty space-time. The cosmological constant permitting expansion arises from spatial features such as these. In theoretical accounts, the early dense inflationary universe has to settle in on this small quantity soon after the big bang, some 10^{120} times lower than the ultra-early density (Rees 2000:97).

Even more interestingly, within conventional big bang cosmology, it has proven to be very difficult to understand why the constant is so tiny. Many different factors contribute to its strength; cosmologists would otherwise expect it to be quite large—so large the universe would fly apart. By some accounts the expected value today is 10^{60} to 10^{120} higher than its tiny life-permitting value. Some mechanism cancels out the contributing factors to nearly but not quite zero. Despite various proposals, at present there is no generally accepted way to derive the minute cosmological constant from particle physics or astrophysics.

Even if such an account is found, the interactions that produce such a minute constant that figures so significantly into the expansion rate, continuing from the explosive big bang, will be remarkable. Martin Rees finds this puzzling:

> Fortunately for us (and very surprisingly to theorists), λ is very small. Otherwise its effect would have stopped galaxies and stars from forming and cosmic evolution would have been stifled before it could even begin. . . . The cosmic number λ—describing the weakest force in nature, as well as the most mysterious—seems to control the universe's expansion and its eventual fate. . . . Our existence requires that λ should not have been too large. (Rees 2000:3, 98–99; cf. Graesser et al. 2004)

Stephen Hawking asks:

> Why is the universe so close to the dividing line between collapsing again and expanding indefinitely? In order to be as close as we are now, the rate of expansion early on had to be chosen fantastically accurately. If the rate of expansion one second after the big bang had been less by one part in 10^{10}, the universe would have collapsed after a few million years. If it had been greater by one part in 10^{10}, the universe would have been essentially empty after a few million years. In neither case would it have lasted long enough for life to develop. Thus one has either to appeal to the anthropic principle or find some physical explanation of why the universe is the way it is. (Hawking 1996:89–90; Penrose 2005:462–464, 772–778)

If we cut the size of the universe from 10^{22} to 10^{11} stars, *then* that much smaller but still galaxy-sized universe might first seem roomy enough, but it would run through its entire cycle of expansion and recontraction in about one year! *If* the matter of the universe were not so relatively homogeneous as it is, *then* large portions of the universe would be so dense that they would already have undergone gravitational collapse. Other portions would be so thin that they could not give birth to galaxies and stars (Wheeler 1975). The universe is immense, but it could hardly be any smaller if life were to appear in it at all.

If the universe were not expanding, *then* it would be too hot to support life. *If* the expansion rate of the universe had been a little faster or slower, *then* connections would have shifted so that the universe would already have recollapsed or so that galaxies and stars could not have formed. The explosive expansion, extent, and age of the universe are not obviously an outlandish extravagance. If it were not vast, lonely, and dark, we would not be here—as we earlier heard John Barrow conclude (Barrow 2002:113). Indeed, this may be the most economical universe in which life and mind can exist—so far as we can cast that question into a testable form in physics.

(2) Astronomical phenomena such as the formation of galaxies, stars, and planets, which depend critically on the expansion rate of the universe, also depend critically on the microphysical phenomena. In

turn, those midrange scales where the known complexity mostly lies depend on the interacting microscopic and astronomical ranges. The human world stands about midway between the infinitesimal and the immense on the natural scale. The size of a planet is near the geometric mean of the size of the known universe and the size of the atom. The mass of a human being is the geometric mean of the mass of Earth and the mass of a proton (Carr and Rees 1979:605). A person contains about 10^{28} atoms, more atoms than there are stars in the universe, and that number puts us mid-scale. "The human scale is, in a numerical sense, poised midway between atoms and stars" (Rees 2001:183).

Planets and persons, at mid-scale ranges, equally depend on the structure and processes at the astronomical and atomic ranges. Change slightly the strengths of any of the binding forces that hold the world together, change critical particle masses and charges, and the stars would burn too quickly or too slowly, or atoms and molecules (including water, carbon, and oxygen) or amino acids (building blocks of life) would not form or remain stable. The charges on the light electron and on the vastly more massive proton are exactly equal numerically. "Heaven knows why they are equal," wondered George Wald, "but if they weren't there would be no galaxies, no stars, no planets—and, worst of all, no physicists" (quoted in *New Scientist* 60, no. 871 [Nov. 8, 1973]: 427). A fractional difference and there would have been nothing.

(3) Those four fundamental forces that hold the world together range over forty orders of magnitude; some involve repulsion as well as attraction, but the push as well as the pull is used to hold things together. The mix of forces is both remarkable and complex.

(3a) Gravity is the weakest of these forces, and only positive. Gravity affects the expansion rate of the universe; without it there would be no galaxies or stars. Since it is the weakest by far, although it does operate at long distances, we might ask why it is so weak. But if gravity were much stronger, the solar system as we know it would not exist. On Earth, most of the structural features of living organisms would be upset (such as bones or brains). Large terrestrial organisms would be impossible. If gravity were much weaker the formation of galaxies and stars would be upset, including the supernovae explosions on which life depends.

(3b) Owing to the electromagnetic force, positively charged protons and negatively charged electrons are attracted. The electromagnetic force is essential to chemistry; think of chemical bonding or the atomic table. Protons repel protons by the same force. John D. Barrow and Joseph Silk calculate that "small changes in the electric charge of the electron would block any kind of chemistry" (Barrow and Silk 1980:128).

(3c) At short ranges, however, the strong nuclear force is stronger and keeps the protons (and neutrons) together in an atom. Without the strong nuclear force there would be no atoms other than hydrogen. If these forces are much changed, basic atomic structures and chemistries are radically altered. When we consider the first seconds of the big bang, writes Bernard Lovell:

> It is an astonishing reflection that at this critical early moment in the history of the universe, all of the hydrogen would have turned into helium if the force of attraction between protons—that is, the nuclei of the hydrogen atoms—had been only a few percent stronger. In the earliest stages of the expansion of the universe, the primeval condensate would have turned into helium. No galaxies, no stars, no life would have emerged. It would have been a universe forever unknowable by living creatures. A remarkable and intimate relationship between man, the fundamental constants of nature and the initial moments of space and time seems to be an inescapable condition of our existence. . . . Human existence is itself entwined with the primeval state of the universe. (Lovell 1975:88, 95)

The value of this strong nuclear force (ϵ) as figured into equations is 0.007. What if it were a little different? asks Martin Rees. His answer: "If ϵ were 0.006 or 0.008, we could not exist" (Rees 2000:2, 48–51).

(3d) The weak nuclear force (a billion times weaker than the strong force) is involved in the relative proportions of protons and neutrons in stars. An aging star may end in a huge explosion, a supernova, which is how the elements it has forged get distributed into space and form planets. First the star collapses and the implosion results, on account of the weak force, in protons turning into neutrons, emitting neutrinos, which, in the dense stellar core, drive the rebounding explosion. These processes would fail if the weak force were much weaker, or stronger. Also, if the weak force were much different there would be only helium stars (with a short lifetime) and no hydrogen stars, in which the heavy elements form.

In this universe at least, these forces, and the particle masses and charges involved, have to be about what they are, if anything more complex is to develop.

(4) Carbon is basic to life as we know it, and there are considerable difficulties in envisioning any alternatives to carbon-based life (Pace 2001; cf. Bains 2004). Likewise, oxygen, required to form water, is vital to life (Finney 2004). Both are produced in stars in abundance by a series of quite precise steps. Hydrogen is converted to helium (helium 4) and the helium subsequently converted to carbon and oxygen. In an intermediate step, helium nuclei collide to form unstable beryllium 8, which is quite short-lived, capturing another helium nucleus to produce carbon (carbon 12). (Beryllium 9 is the stable form, a light metal.) Some of the carbon collides with helium nuclei to produce oxygen (oxygen 16) (Clayton 1983). These transformations are rather surprisingly adjusted so that abundant amounts of both carbon and oxygen are produced. If not, there would be mostly carbon or mostly oxygen.

Fred Hoyle, an astronomer, was quite startled by his discovery of these critical levels through which carbon just manages to form and then only just avoids complete conversion into oxygen. If the strong force had varied by a half a percent, the ratio of carbon to oxygen would have shifted so as to make life impossible.

> Would you not say to yourself . . . "Some supercalculating intellect must have designed the properties of the carbon atom, otherwise the chance of my finding such an atom through the blind forces of nature would be utterly minuscule"? Of course you would. . . . You would conclude that the carbon atom is a fix. . . . A common sense interpretation of the facts suggests that a superintellect has monkeyed with the physics, as well as with chemistry and biology, and that there are no blind forces worth speaking about in

> nature. The numbers one calculates from the facts seem to me so overwhelming as to put this conclusion almost beyond question. (Hoyle 1981:12)

These synthesis processes involve what are called "resonance states" in carbon and oxygen, which depend on the strong nuclear force and the electromagnetic force. Compare how a radio has to be tuned right on frequency to get the desired station—the origin of the "fine-tuned" metaphor. Astrophysicists have made quantitative analyses of the effects of changes in these forces on the amount of carbon (C) and oxygen (O) produced. "We conclude that a change of more than 0.5% in the strength of the strong interaction or more than 4% change in the strength of the Coulomb force [of electromagnetic attraction and repulsion] would destroy nearly all C or all O in every star" (Oberhummer et al. 2000:90).

The six elements especially important for life are hydrogen, carbon, oxygen, nitrogen, sulphur, and phosphorus. Hydrogen is everywhere, and of the heavier elements, four are produced in relative abundance. But phosphorus (phosphorus 15), quite vital to life, is not common, because its synthesis requires many complex nuclear reactions, which can only take place in a minor subset of massive stars, and these produce little (Maciá et al. 1997). Though uncommon on Earth, phosphorus is, fortunately, concentrated in minerals such as apatite, from which it, with its distinctive chemical properties, becomes available in sufficient amounts for the energies and structures of life. In the form of phosphates, conversions from ATP to ADP are the energy currency of life. Phosphate groups with sugars form the backbone links in DNA and RNA; phosphates are important in the synthesis of biological molecules (such as phospholipids in lipid bilayers) and in vertebrate skeletons. As part of the molecule NADP, phosphorus is essential in electron transport and oxidation-reduction reactions (Westheimer 1987; Williams 2000; Skinner 2002).

There are as well several metallic ions required, often in only trace amounts, for key metabolic processes—magnesium, iron, and zinc, in the structures of chlorophyll, cytochromes, hemoglobins, and some enzymes (Williams 1953, 2000). These elements too are sufficiently available. So in sum, the nucleosynthesis of the biogenic elements, with their remarkable structural and metabolic possibilities, is both fortunate and impressive. As a result of the stellar physics and chemistry, the building blocks are in place. The carbon-oxygen synthesis seems exactly and singularly fine-tuned, but other syntheses are more plural and diverse, cooking up about a hundred elements, with varied nuclei and electron shells and subshells, those grappling hooks with such immense possibilities for constructions. A dozen of these elements have possibilities that are biologically tantalizing. Still, in the nucleosynthesis in the heavens above, there is not a shred of any cybernetic know-how to engineer the much more complex hookups that launch and sustain life. That came about in the second big bang on Earth.

(5) The parameters of physics contain a small constant called the fine structure constant, usually designated α, which controls the strength of the electromagnetic force. The constant, which is a dimensionless quality, is $7.297352570 \times 10^{-3}$, or

$$\frac{1}{137.035999070}$$

It was first discovered during analysis of the fine structure of atomic energy spectra, but has proved to be important because it sets the relation of the electromagnetic force to the strong nuclear force, about a hundred times weaker. The strong nuclear force holds a multiproton nucleus together, despite the electromagnetic force (proton repelling proton) that would blow it apart. Cosmologists debate whether the fine structure constant might be quite slowly changing, but if it had long been much different, that would have changed the number of elements that can exist in nature, shifting these force relationships. If it were much stronger, many fewer elements could exist, including many necessary for life—if 4 percent stronger, carbon would no longer be produced in stellar fusion. If stronger, this would also change chemical bonding, which occurs on the basis of properties of the electrons in outer shells. Thus the bonding and folding protein chemistry on which life depends would not be possible (Barr 2003:125–126; Barrow 2004:411–419).

(6) Physicists speak of symmetry-breaking (Brading and Castellani 2003). The term is widely used in various sciences when a lawlike system moves from a prior state with as yet some undifferentiated order into a differentiating juncture, where the developing system splits, transforms, takes one direction and forgoes others. The process crosses a critical point where spontaneous passage events, perhaps small fluctuations, determine which branch is taken, subsequently operating under laws differentiated in the now broken symmetry. With objects, a hydrogen atom is symmetrical; so is an oxygen atom; but when joined as H_2O the molecule is polar, with positive and negative ends, and thus asymmetrical.

Homogeneous structures may have great symmetry (oxygen atoms, salt crystals), but complex structures, if preserving some symmetries, also require much asymmetry (a DNA molecule, a spiral helix with a unique linear triplet sequence). There is, paradoxically, less order of the symmetrical kind to achieve more order of the complicated kind—which may also involve some messiness. Breaking symmetry may be the first step to getting something happening. There is phase change with novelty—with "broken symmetry," as F. W. Anderson put it in a famous paper, "More is Different" (Anderson 1972).

Cosmologists suppose symmetry-breaking at the startup of the universe. In big bang cosmology, despite worries that we earlier noticed about whether scientists have rational access to the first moments, some suppose (speculate) that during high temperature Planck time (the first 10^{-43} seconds) the four forces were unified into a single superforce. Afterward, with lowering temperatures, there were the earliest symmetry breaks that separated the original force, resulting in the four forces we know in the present universe (fig. 1.1, Nave 2002). Cosmologists may pay particular attention to the breaking of the once-combined electromagnetic-weak (electroweak) force into the electromagnetic and weak forces. The single superforce explodes into "more," and that is "different." Something starts to happen.

These different forces, as already noted, are responsible for features of stars and planets, the diversity of the elements, and the bonding

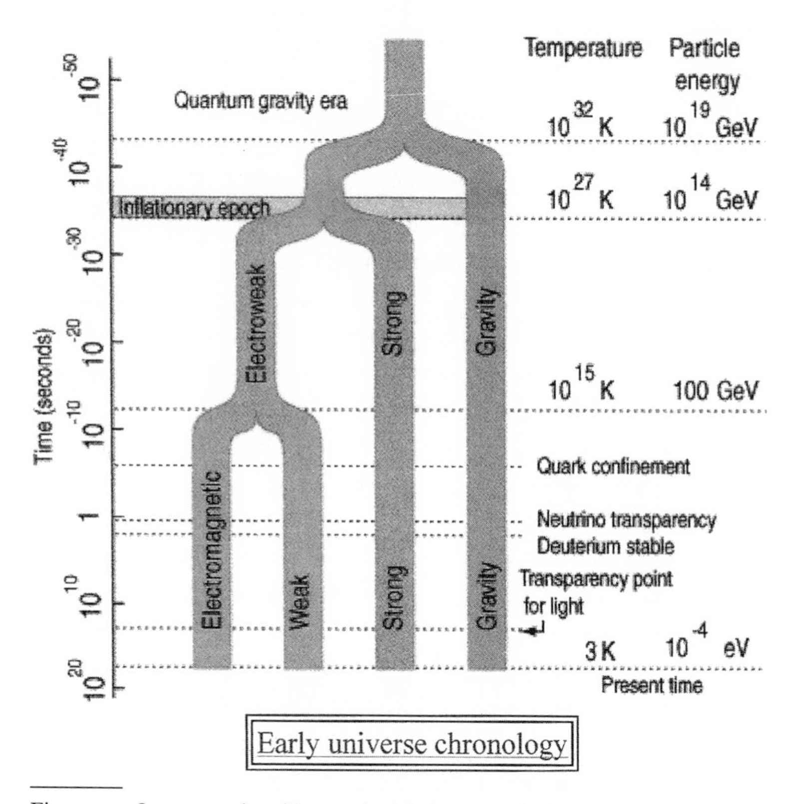

Figure 1.1 Symmetry-breaking at the big bang explosion (*Nave 2002*)

chemistries without which biomolecules and life would not be possible (making water possible, for instance). John Barrow remarks: "The situation in which the outcomes of a law break its symmetry is termed 'symmetry-breaking.' It has been known but not fully appreciated for years. And it is responsible for the vast diversity and complexity of the real world" (Barrow 2007:138). David Gross, in a review of symmetry in physics, concludes: "The secret of nature is symmetry, but much of the texture of the world is due to mechanisms of symmetry breaking" (Gross 1996:14257). Accounts of symmetry-breaking during the primordial big bang are speculative and likely to change. But somehow or other these four forces did come into being in their present biocentric/anthropic form. If they had not, the universe we know, life in it, and we ourselves would not be here. That is not speculative.

Cosmic Results: Predictable and Surprising

Are these ongoing results of the primordial big bang predictable or surprising? Or both? Is the outcome prespecified? Blueprinted into the universe—this universe at least, as a theory of everything might suggest? Or is the outcome in significant part contingent, and surprising, as Davies' "cosmic jackpot" suggests?

"Prediction" and "surprise" have been analyzed in philosophies of science. Science prefers lawlike regularities, which, when applied to a set of initial conditions, enable prediction. There are no surprises. But science often finds surprises, indeed dramatic singularities, in natural history. Typically, this forces the question whether with a better law, applied to those initial conditions, the outcome is no longer surprising but rather to be expected. At the cosmic level, scientists have no such theory of everything. When philosophers or theologians wonder what to make of this, they may say that some deeper explanations are needed, such as God or Platonic forms, whereupon the jackpot universe becomes less surprising.

However, if there is discovered a theory that makes the three big bangs predictable, they may again say that we need a deeper account of why there should be a universe with such built-in inevitability. Either way, in a surprising universe or an inevitable universe, we need an account of matter-energy generating life, of life generating mind.

It is difficult to envision any cosmology that does not require creation of the complex out of the simple, more out of less, something somehow out of nothing. It is difficult to imagine that all of the remarkable phenomena that have worked together to make our universe possible will disappear. It is difficult to imagine a universe more staggering, dramatic, and mysterious, for all its rationality. It is difficult to imagine a universe that starts simpler (perhaps as quantum fluctuation in a vacuum) and becomes more complex (*Homo sapiens* sequencing its own genome, with moral debates whether and how we ought to revise ourselves). The universe story, the Earth story is a phenomenal tale of more and more later on out of less and less earlier on. As events move from quarks to protons, from amino acids to protozoans, to trilobites, to dinosaurs, to persons, from spinning electrons to sentient animals, from suffering beasts to sinful persons, the tale gets taller and taller. No doubt there will be surprises in cosmology in the next century; surprises in discovering biodiversity, terrestrial, even extraterrestrial; and surprises in human achievements and powers, a taller tale still. Is the end of the story somehow already there in the beginning?

Generating these heavy elements, which on Earth become the seeds of life, does seem deterministic in origin. In that sense the periodic table of chemical elements is latent in the big bang—including those remarkable biogenic elements. So are the thirty-two crystal classes. Molecular structures, molecules and lattices, as found in water, pyrite, salt, and silica, inevitably develop somewhere. The system is prone to modular constructions, which may get intertwined or compounded (hypercycles), and the stable and metastable ones survive. Random elements combine with overall order (as with fractals). Beyond aggregation, matter is regularly spontaneously organizing, as when molecules and crystals form. In some situations, especially with a high flow of energy over matter, patterns may be produced at larger scales (Prigogine and Stengers 1984). These patterns may further involve critical thresholds, often called self-organized criticality (Bak 1997). Such processes are "automatic," sometimes called "self-organizing," although initially the "auto" should not be taken to posit a "self," but rather an innate principle of the spontaneous origination of order.

Planets may form with differences, especially if there are chaotic factors in their origins. "According to modern views, the number of planets and the size of planetary orbits was determined more or less accidentally during the complicated process whereby our solar system condensed out of a gigantic interstellar cloud" (Wilcsek 1999:303). But where planets do form, given suitable conditions on some of them, igneous, sedimentary, and metamorphic rocks are inevitable. Where there are fluids, there will appear loops and cycles, bubbles, currents, eddies, tornadoes. There seem to be chaotic thresholds that trigger amplified particulars (such as the Grand Canyon), but there are canyons generically—on the moon, Mars, and elsewhere in space.

There are other planets. The presence of several hundred possible planets has been detected, and these do seem to be diverse, though none suitable for life is yet known. Rather, it seems that planetary systems configured like our solar system are quite rare. If there proves to be a second (or prior) genesis of life elsewhere, that will be welcome. But Earth will not on that account cease to be remarkable (a remarkable accident?), nor will its particular natural history—trilobites, dinosaurs, primates—and social history—Israel, Europe, China, global capitalism, the Internet—cease to be unique in the universe.

We could make better estimates of the random and the probable if ever we did discover a second genesis (astrobiology, exobiology). We might yet discover life in our solar system, though discovering intelligent life seems quite unlikely. We might detect life on relatively near stars, but about life in other galaxies we are likely to remain long in ignorance; the distances are too great. The only second genesis we are likely ever to be able to detect in the vast reaches of space is the unlikeliest of all that we will find: intelligent life smart enough to transmit electronic signals across space. Peter D. Ward and Donald Brownlee make such conclusions, indicated in their title: *Rare Earth: Why Complex Life Is Uncommon in the Universe* (2000). Again, given the distances in space and time, we are unlikely ever to be able to communicate with such life. We may be the result of cosmic natural history generally, of earthen natural history peculiarly, the most complex event in the universe, and stuck in our solitude.

A good planet is hard to find, and Earth is something of an anomaly, so far as we yet know. Earth has a rather good star, the sun, which is stable, solitary, and situated about 28,000 light-years from the center of our galaxy, a relatively quiet part of the galaxy, about halfway between the quite active middle and the active outer parts. Deadly radiation from supernovae explosions or bursts of intense X-ray and ultraviolet radiation are unlikely. The solar neighborhood does have a relatively high abundance of the heavier elements produced from supernovae, all those heavier than hydrogen and helium—an abundance the astronomers call "metallicity."

Most planets, even though they contain suitable elements, will not be in a habitable temperature zone. Located at a felicitous distance from the sun, Earth has huge amounts of liquid water: seven oceans covering about three quarters of its surface. "Aqua" would have been a better name than "Earth." Water is an anomalous liquid, with a maximum density near 4 degrees Celsius, so that ice floats, with high latent heat and slow evaporation, with solvent properties suitable for life. The polar structure of water and its anomalous properties later fit impressively into the fundamental support of critical biological structures in carbon-based life (Finney 2004). These include lipid bilayers, which organize the compartments within cells, or the funneling and maintaining protein folding (Chaplin 2001; Levy and Onuchic 2006). Such properties depend on the strengths of those bonding forces noted earlier; they are inevitable, but become available for life only on a lucky planet such as Earth/Aqua.

On Earth there is atmosphere, a suitable mix of elements, compounds, minerals, and an ample supply of energy. Radioactivity deep within the Earth produces enough heat to keep the tectonic plates of its crust constantly mobile in counteraction with erosional forces, and the interplay of such forces generates and regenerates landscapes and seas—mountains, canyons, rivers, plains, islands, volcanoes, estuaries, continental shelves. Earth's moon produces tides, significant in the evolution of life. "It appears that Earth got it just right," conclude Ward and Brownlee (2000:265). All this results in an anomalous, fortuitously good location—though it is hard to say whether the jackpot Earth is lucky, likely, or inevitable. William C. Burger does call Earth a "perfect planet." "I believe we can all agree that we live on a glorious planet, and that our intellectual achievements have been quite amazing" (Burger 2003:3). On Earth, complexity does increase again, by many more orders of magnitude.

Events resulting from the big bang might be contingent at several points:

(1) The setup at the time of the big bang might have been otherwise; it was what it was owing to startup contingencies. We examined this issue earlier, wondering if there might be a "theory of everything" that would determine these events, making it inevitable, if a big bang happened, that it happen the way it did, with the fine-tuned results. No such theory exists at present. Even if it did, one would still have to wonder whether such a determinate process had to be launched, and also to puzzle about how the requirements of physical theory happened to be also those required for life on Earth.

(2) By another account our universe can be contingent within some ensemble of universes, one of multiple universes, and a lucky one. Currently the most fashionable way to account for the surprising aspects of our universe, our big bang, is to suppose that there was not just one big bang but very many. Or perhaps our one big bang produced a megaverse with regional, disjoint universes, bubble universes, each different—a bigger big bang than we had thought and bigger than we can observe from our location. There is no one universe but multiverses, multiple universes. Astronomy has steadily been increasing the size of this universe, and now we need to suppose more vastness again.. This universe is only one of a run of universes (Carr 2007; Rees 2001). There are other, parallel space-times with different laws, constants, contents, histories. Most are uninteresting, but some lucky ones are interesting. Most are not cognizable worlds because their laws and constants are such that those kind of worlds cannot produce cognizers who inhabit them. Maybe even all the universes that can exist do exist (Tegmark 2003). If so, it is not so surprising that some of them are suitable for the formation of heavy elements and the evolution of life.

Another possibility is: big bang, big squeeze, big bang, big squeeze, a recycling universe, sequential universes, and our particular current cycle at random has the right characteristics for life (Wheeler 1975). This, however, looks unlikely with recent evidence that our universe will continue expanding forever. By another account, black holes are generating other universes (Smolin 1997). If there are multiple universes, what is unlikely on one throw of the dice becomes likely if you roll the dice often enough.

But these are complex explanations indeed—to invent myriads of other worlds existing sequentially or simultaneously with ours, in order to explain how this one can be a random one from an ensemble of universes, and so a little less surprising in its anthropic features. There is of course little or no scientific evidence that other universes exist. They are mostly scientific conjectures, if indeed these are scientific hypotheses of the familiar kind at all, since the existence of other, nonobservable universes with differing constants and laws is not directly testable, only marginally so by extrapolation from what we may know about this one.

Their existence might make some phenomenon in our universe more plausible. But we in our universe are too separated from them to have a science about them.

Their proponents seem inclined to believe they exist not so much on the basis of any actual evidence that these other universe-regions exist, but rather because, if true, that would make this universe less surprising. Cosmologists of course are licensed to speculate (Tegmark 2003). True, this universe we do inhabit has often proved immensely bigger than we supposed, but we have strong scientific evidence for other galaxies, for deep space, deep time. Martin Gardner puts the contrast bluntly: "The stark truth is that there is not the slightest shred of reliable evidence that there is any universe other than the one we are in. No multiverse theory has so far provided a prediction than can be tested" (Gardner 2003:9). So in the meanwhile, it seems a more economical explanation (remembering that science often urges simplicity in explanation) to posit only the one universe we know and some constraints on it that make it right for life.

(3) Given some startup big bang, contingent or inevitable, there might also be contingency en route in the unfolding natural history. The basic laws and constants might (after the startup) be determinate, but there might be contingency nevertheless within the framework of such basic laws and constants. The laws include quantum physics, for example, which has indeterminacy within it, by most accounts an ontological indeterminacy. If quantum events can ever be amplified to larger scales, those results would be to some degree contingent.

In fact, we have not far to seek for evidence that molecular and even atomic phenomena are often amplified. In biochemistry and genetics, events at the phenotypic level are profoundly affected by events launched at the genotypic level. Such events may sometimes be affected by quantum events, as when random radiation affects point mutations or genetic crossing over. This may affect enzyme functions or regulatory molecules, as when allosteric enzymes, which amplify processes a million times, are in turn regulated by modifier molecules, of which there may be only a few copies in a cell, made from a short stretch of DNA, where a few atomic changes can have a dramatic real-life effect. A single base pair altered can shift a whole reading frame.

If radioactive decay caused a mutation that altered efficiency in photosynthesis and conveyed survival advantage, that would affect events at ecosystem scales. Indeed, by the usual evolutionary account, the entire biological tale is an amplification of increments, where microscopic mutations are edited by macroscopic selective processes. These increments are most finely resolved into molecular evolutions.

Mostly, quantum indeterminacies wash out at our native range levels. That is required for the order of natural law. A macrodeterminism remains, despite a microindeterminism. The physical world is in fact routinely described in statistical terms. This is often because of our epistemological uncertainty, incomplete information. But objectively random processes at one level can yield reliable results at other levels; the random distribution of grains of clay in a brick nevertheless permits a stable and ordered wall in a building. Despite the atomic uncertainties, we can still have clocks accurate to millionths of a second, because the averages are that reliable. Stochastic processes at lower levels are compatible with determinate processes at upper levels. If this is all there is to be said, the atomic indeterminacies imply nothing for human affairs or for a broad-scope worldview.

But in statistical systems with chaotic elements, some of them genuinely indeterminate, random differences at a threshold during initiation can lead to widely different outcomes. Whether a fire starts when a spark falls into a dry forest depends on coincidences at the start—a few drops of rain, a puff of wind, how a few fallen leaves happen to lie—although once the flame is ignited, spread of the fire becomes nearly certain. Even on a global scale, climatologists now allow that weather systems, even climate systems, have indeterminate dimensions (Lorenz 1968). Given billions of years of natural history, it seems likely that at times and places, mutations bubbling up from atomic indeterminancies have resulted in important shifts in adapted fit.

(4) The contingencies at physical levels, present as ongoing openness and indeterminacy, seem also to be subject to a novel biological "top-down" causation. If we turn from the *random* to the *interaction* possibilities in physics, we gain a complementary picture. Physical nature, as it resulted from the explosive big bang, is not just indeterminate in random ways; when organisms arrive at the second big

bang, biological nature takes advantage of physical nature. We gain space for the higher biological phenomena that physics leaves out, yet for the possibility of which physics provides. In the chapter to come, we will discover that microscopic indeterminism provides a looseness through which an organism can steer itself by taking advantage of the fluctuations at the micro levels, resulting in an explosion of biodiversity and biocomplexity. This provides a different perspective on the mixing of order and disorder. Complexity requires multiple distinct parts with multiple connections. Too much distinctness yields disorder, chaos, contingency. Too much connection yields rigidity, determinism, order. Complexity must be situated between order and disorder.

Systemically, there seems a mixture of inevitability and openness, so that one way or another, given the conditions and constants of physics and chemistry, together with the biased earthen environment, life will somehow both surely and surprisingly appear. Manfred Eigen, a thermodynamicist, concludes "that the evolution of life . . . must be considered an *inevitable* process despite its indeterminate course" (Eigen 1971:519). Life is destined to come, yet the exact routes it will take are open and subject to historical vicissitudes. Others, although they may agree about the openness, are not so sure about the inevitability.

Our particular universe is "singular," as we have recognized, but then it develops nomothetic order, the laws of physics, chemistry, geology. But nothing in these laws demands that this proper-named planet, Earth, be produced. Certainly, anticipating our second big bang, life on Earth was a unique singularity, which hardly seems to have been "front-loaded" in any astrophysics, microphysics, chemistry, geology. There is no way that the most learned extraterrestrials visiting our galaxy, arriving at our solar system, flying by Saturn, Jupiter, and Mars, could predict the elephants they will find on Earth. All this driven/attracted behavior of matter-energy contrasts with a more mixed, open account in biology, to which we next come.

So one of the surprises of contemporary physics is that the human person is composed of stardust, fossil stardust! Or if you prefer to be more dismal about being lost in the stars, we humans are leftover nuclear waste. What should we make of this? Sometimes we dismiss the puzzle, noticing that in no other kind of universe could humans have evolved to worry about these things. We are here and it really is not surprising that the universe is of such kind as has produced us. We knew before we started our search that the universe has all the prerequisites for our being here. The anthropic principle is an observer selection effect, something like being surprised that all your ancestors survived long enough to reproduce, human and prehuman back to the startup of life. But those who want a fuller explanation will find it quite impressive to discover that what seem to be widely varied facts really cannot vary widely, indeed, that many of them can hardly vary at all, and have the universe develop the matter, life, and mind it has generated.

Stephen M. Barr, as a theoretical particle physicist, comments that physicists

> cannot get around the fact that our universe is a special kind of place—indeed, doubly special. . . . It is a very curious circumstance that materialists, in an effort to avoid what Laplace called the unnecessary hypothesis of God, are frequently driven to hypothesize the existence of an infinity of unobservable entities [such as other universes]. . . . It seems that to abolish one unobservable God, it takes an infinite number of unobservable substitutes. (Barr 2003:156–157)

Perhaps an answer to whether these ongoing results of the big bang are predictable or surprising is to ask whether, from the beginning, the possibilities were already there. William R. Stoeger, both astronomer and theologian, considers this from a theological perspective.

> God, as Creator, endows nature from the beginning with existence and with capacities and dynamisms to evolve the rich diversity of remarkable structures and organisms which have emerged in the course of cosmic history. Included with this endowment is relative freedom and autonomy—the course of evolution was not rigidly determined from the beginning, but the rich potentialities were there. Some of these were realized and others were not. . . . God's direct intervention in the evolutionary process as another secondary cause is not needed. (Stoeger 2007: 242–243)

So no possibilities emerge en route; the potentialities were there at the big bang and have been since unfolding, with some openness in the unfolding.

John Polkinghorne, physicist and theologian, with his colleague Nicholas Beale, puts it this way:

> The universe started in an extremely simple way. Following the big bang it was just an expanding ball of energy. Now, after 13.7 billion years, it is rich and complex, the home of saints and scientists. . . . As we have come to understand many of the processes by which this great fertility has come about, we have come to see that their possibility had to be built into the given physical fabric of the world from the start. (Polkinghorne and Beale 2009:13)

There is no enlarging possibility space. The later singularities were always lurking around, though concealed, from the startup. Nothing ever becomes explicit that was not forever latent. The stars in intense heat are disposed to form sodium and chlorine and, in cooler environments, sodium and chlorine are disposed to form salt. There can be endless openness within structured frameworks.

Water was already there in the possibility spaces of hydrogen and oxygen, before any water molecule had ever formed, in the sense that if ever two atoms of hydrogen met one of oxygen under certain physical conditions, they would spontaneously join to form water. But DNA molecules coded for making hemoglobin are not already there all along in a soup of the relevant atoms (hydrogen, oxygen, nitrogen, phosphorus), any more than hemoglobin (an allosteric protein, shifting its shape to carry oxygen in blood, built from four myoglobin molecules) is already there in a pile of its atoms before there is any DNA. The relevant information introduces new possibilities not previously there.

Suppose that a meteorite lands on Earth, releasing some iron atoms as the incandescent meteor crashes into the ground. Suppose some of those iron atoms make their way into my diet, and into my blood. Would not such meteoric iron, from outer space, work just as well as any terrestrial iron atom carrying oxygen to my brain? Does that not mean that such iron atoms have had from time immemorial the capacity for entering into cognitive processes? Passively perhaps, if overtaken by mind, but actively there is no such self-contained potential. A single atom of iron has no such possibilities within itself at all. To claim that it does is like saying that ink and paper have the all the possibilities of the Library of Congress latent within the bottle and secretly coded in the paper pulp fibers. Entering into thinking processes becomes a possibility for such an extraterrestrial iron atom only upon its encounter with (only relative to) the systemic company of enormous amounts of information.

One can insist that it must always have been possible to put carbon atoms into organic cells and silicon atoms into computers, since we humans do that now somatically and technologically—and the atoms are no different from what they have been for billions of years. But it may have always been possible to do this *with* these atoms, provided that one had the know-how to do such things, but not possible lacking such information. Such information has to become possible. That is a different claim from the claim that it has always been possible *for* carbon and silicon to self-organize into organism and computers.

We know that water, as a polar molecule, has various features that have turned out to be fortunate for supporting life. But you can know all about the polarity of water, and nothing known there leads you to predict lipid bilayers later on, built with their hydrophobic heads and hydrophilic tails, used to make membranes that enclose the life structures. In the forest, a scientist encounters a tree, the wood functioning to hold the leaves up to the sun. But what new can we do with wood? We can build a violin and play music. This gives us no cause to claim that a violin is lurking in the possibility space of the tree with its wood.

A kaleidoscope always has the possibilities for the patterns it produces, though the exact patterns that become actual also result from the random tumble of the bits of colored glass within it. Snowflakes form in endless variety, but all are frozen into a six-sided crystalline latticework, presumably latent in physical crystallography from the beginning. But does this also mean that the protons, electrons, and atoms were waiting to string themselves together in DNA, waiting for a context of sufficient entanglement to form vital nodes in networks, the metabolisms of life? Kaleidoscopes and snowflakes do not go anywhere; but DNA does, from zero to billions of species, humans included. Is all this latent from the start? Were the rudiments of life always there? Or is there breakthrough? Is there any nudging?

Stoeger does not find or want any divine intervention. God does everything primordially and pervasively yet not much of anything specifically. But Stoeger soon adds:

> God not only creates the universe from nothing, but also holds it in existence at each moment. And God not only holds it in existence at each moment, but is also working and struggling as Creator through the laws of nature, and the processes of cosmic, biological, and social evolution, to coax it towards the realization of its destiny. (Stoeger 2007:244)

There seem to be no new potentialities, no intervention, and yet God simultaneously working, struggling, coaxing to make some potentialities actual and to avoid others. God is "actively engaged with and supportive of the emergent capacities (such as personhood) at each level" (Stoeger 2007:247). So does Stoeger think life and mind are front-loaded into matter? Potentially yes, it seems, but not likely to emerge without God's active (if nonintervening) coaxing.

Polkinghorne and Beale similarly need guidance amid the possibilities:

> The creation has been endowed with great potentiality (remember fine-tuning), but the manner in which that potentiality has been brought to birth in particular ways is through the shuffling explorations of the evolutionary process. The history of the universe is not the performance of a fixed score, written by God in eternity and inexorably performed by the creatures, but it is a grand improvisation in which the Creator and the creatures cooperate in the unfolding of the grand fugue of creation. . . . God can bring about determinate ends, even if they are achieved along contingent paths. (Polkinghorne and Beale 2009:15)

So even if these authors hold that there is no enlarging possibility space, they do think that the outcome is not inevitable—short of divine cooperation, co-operation. Using our terms, the first big bang is necessary, but not sufficient for the second and the third. The first big bang makes the other two possible—but if and only if there is some coaxing, some innovation. That sounds, though, as if something more does have to be added to produce the further results.

There is a tension between two ways of speaking of possibilities. By one account, all the possibilities are there at the start, an all-but-infinite library of possibilities. The big bang is superfertile with possibilities. Each realization of actual events (the formation of the Crab Nebula, or of Saturn) realizes one possibility from among many and thereby shuts down previous possibilities that earlier existed in this immense possibility space (something like marrying one woman or choosing one career shuts down other marriages and careers). Across the 13.7 billion years, the possibilities have been getting steadily fewer with each entity actualized, with any process completed.

But it seems equally plausible to argue that new possibility spaces do open up that were not there at the start. Life is not possible on Saturn, but it does become possible on Earth. On Earth, it is not possible for trilobites to build jet planes, but that does become possible for humans. We will next be developing how with genetics, much becomes possible that was not previously possible. Especially in the explosions of biology on Earth, new possibilities seem to open up, dramatically, overwhelming with their increase any old ones shut down.

Polkinghorne and Beale concede this, when they add "active information" to the creative process. "New causal principles come into play. . . . The behavior of such systems is no longer completely determined by the 'low-level' laws and adds a logic of its own." "Every so often in the history of the universe something intrinsically new emerges from within the deep potentiality with which creation was endowed" (Polkinghorne and Beale 2009:17, 51, 81). That makes the difference between what is necessary and what is sufficient for the sequential three big bangs. The possibility space at the startup permits, but does not require enlarging possibility space to emerge. Natural history happens forward, from start to finish, but narratives have to be understood retrospectively, the beginnings in the light of the endings. In great stories, past, present, and future are in reciprocal illumination.

These inquiries in recent decades about the character and results of the primordial exploding contrast strongly to the universe portrayed a century back, before quantum physics, relativity theory, and the advent of contemporary physics. In previous centuries, physics seemed to find that we humans, on a small planet, inhabited a fantastically huge clockwork, rockwork universe, a mechanism of matter in energetic motion. The heavens were fabulously big but no longer seemed heavenly. There was celestial decay. In the sky was more dirt, mountains on the moon, asteroids in deep space. Humans were cosmic dwarfs, lost out there in the stars.

> The world that people had thought themselves living in—a world rich with colour and sound, redolent with fragrance, filled with gladness, love and beauty, speaking everywhere of purposive harmony and creative ideals—was now crowded into minute corners in the brains of scattered organic beings. The really important world outside was a world hard, cold, colourless, silent, and dead; a world of quantity, a world of mathematically computable motions in mechanical regularity. (Burtt 1952, 1996:238–239)

Karl Jaspers found Earth "a minute grain of dust in the universe . . . in an out-of-the-way corner. . . . The fundamental fact of our existence is that we appear to be isolated in the cosmos. We are the only articulate rational beings in the silence of the universe. . . . This is the place, a mote in the immensity of the cosmos, at which being has awakened with man" (Jaspers 1953:237). He hoped that, still, this waking up might be authentic and significant. But when Steven Weinberg got back to the first three minutes, he concluded famously: "The more the universe seems comprehensible, the more it also seems pointless" (Weinberg 1988:154). Weinberg does find impressive rationality, but seems unable to ask whether the increasing comprehensibility he and other astrophysicists are detecting is pointing anywhere.

Roger Penrose is impressed by "the extraordinary degree of precision or 'fine-tuning' for a Big Bang of the nature that we appear to observe" He concludes that ours is an "extraordinarily special Big Bang" (Penrose 2005:726, 762). Martin Rees concludes: "We should surely probe deeper, and ask why a unique recipe for the physical world should permit consequences as interesting as those we see around us" (Rees 2001:163). The startup looks like a setup.

CHAPTER 2

Life

Earth's Big Bang

The most spectacular thing about planet Earth, says Richard Dawkins, is an "information explosion," even more remarkable than a supernova among the stars (1995:145). Nature on Earth has spun quite a story, going from zero through several billion species, evolving microbes into persons. M. J. Benton concludes: "Analysis of the fossil record of microbes, algae, fungi, protists, plants, and animals shows that the diversity of both marine and continental life increased exponentially since the end of the Precambrian" (Benton 1995). Steven M. Stanley agrees: "The empirical record of diversification for marine animal life since Paleozoic time represents actual exponential increase" (Stanley 2007:1). Geerat J. Vermeij finds that "escalation characterizes the Phanerozoic history of life" (Vermeij 1987:419). Andrew H. Knoll celebrates "Earth's immense evolutionary epic": "The scientific account of life's long history abounds in both narrative verve and mystery" (Knoll 2003:1).

Proactive Genetic Information and Order

How are we to understand this second big bang? Resulting from the first big bang, there were two metaphysical fundamentals: matter and energy.

Einstein reduced these two to one: matter-energy. In the rapidly expanding universe, there is conservation of matter, also of energy; neither can be created or destroyed, although each can take diverse forms, and one can be transformed into the other. In the biological big bang, the novelty is that matter-energy enters into information states. The biologists also claim two metaphysical fundamentals: matter-energy and information. The latter is radically novel: proactive information about how to compose, maintain, communicate, and elaborate vital structures and processes. This is information about directed use, which was not present in the previous physico-chemical results of the first big bang.

The information is coded in DNA, and we do not know how DNA-coded life originated. Various scenarios have been proposed, such as the earliest life being RNA based. But since we do not know how life originated on Earth, it is difficult to say whether the second big bang was unlikely, probable, or inevitable. Most scientists think quite unlikely. George M. Whitesides, a chemist, asks:

> How remarkable is life? The answer is: *very*. Those of us who deal in networks of chemical reactions know of nothing like it. . . . How could it be that any cell, even one simpler than the simplest than we know, emerged from the tangle of accidental reactions occurring in the molecular sludge that covered the prebiotic earth? We . . . do not understand. It is not impossible, but it seems very, very improbable. (Whitesides 2008:xiii, xvii)

Francis Crick agrees: "In some sense, the origin of life appears at the moment to be almost a miracle, so many are the conditions which would have had to be satisfied to get it going" (Crick 1981:88).

Because these events are molecular and so small scale, and also because this may be called "primitive" life, we may fail to recognize how dramatic and complex these steps were. Further, at the startup there was no natural selection in any Darwinian sense, since this requires a population of individuals who are reproducing with heritable variation, competing for resources, with differential adapted fit, resulting in differential survival (Gabora 2006). Coded, proactive, environmentally adapted reproduction cannot be used to start up itself, although once started, it can elaborate itself. Lynn Margulis puts this pointedly: "To go from a bacterium to people is less of a step than to go from a mixture of amino acids to that bacterium" (quoted in Horgan 1996:140–141).

Finding a second genesis of life on another planet, or repeated genesis elsewhere, would make life seem more likely, even if we did not know the details of its origins. In developing genetics, such information escalates, crossing critical thresholds. We will return to the question whether, once life is launched, there will be headings or tendencies.

Now there appears a new type of order. A crystal is ordered (formed) spontaneously. There is repeated spontaneous structure formation. A protein molecule is ordered because it is "ordered" to form under the "informed" direction of a DNA molecule, switched on by the organism with its needs. The various spontaneously assembled phenomena in physics and chemistry, for example those called dissipative structures (such as Bénard cells that form in liquids with high temperature gradients) have a physical order but nowhere approach this biological sense of order. Nothing is transmitted from one generation of Bénard cells to the next. In similar circumstances such cells generate again, but they do not regenerate. There is no increasing complexity in the course of reproduction. Similarly with what are called "biomorphs," crystalline structures that resemble biological forms such as curled leaves or worms (García-Ruiz et al. 2009; Kunz and Kellermeier 2009). To be alive requires bounded localized modular assembly (a cell, cells) that continuously regenerates itself (metabolism), replicates itself (reproduction), and is capable of evolving.

Two decades ago what needed to be explained was the generation of *complexity*. In recent decades scientists have come to focus more on the *information* required for specifying and generating such complexity. Norbert Wiener, a founder of cybernetics, insisted: "Information is information, not matter or energy" (Wiener 1948:155). The physical world is composed of matter and energy, with the two united in relativity theory—so physics and chemistry have insisted. But the earthen world, biologists now insist, is composed by information that superintends the uses of matter and energy.

This biological sense of information is proactive, agentive. Such vital information is carried in the genes. What makes the critical difference is not the matter, not the energy, necessary though they are; what makes the critical difference is the information breakthrough with resulting capacity for agency, for *doing* something. Something can be discovered, learned, conserved, reproduced on Earth, but not on the moon. Afterward, as before, there are no causal gaps from the viewpoint of physicist or chemist, but there is something more: novel information that makes possible the achievement of increasing order, maintained out of the disorder. The same energy budget can be put to very different historical uses, depending on the information in the system. What makes the critical difference is not the chemistry, necessary though this is; it is that *reagents* become *agents*.

In Earth's big bang, singularly different from the primordial big bang, nature wonderfully, surprisingly, regularly breaks through to new discoveries because there is new proactive information emergent in the life codings. These achievements are, if you like, fully natural—they are not unnatural; they do not violate nature. But they also are novel achievements of "know-how," of agentive power. Something higher is reached, something "super" to the precedents, something superimposed, superintending, supervening on what went before; there is more where once there was less. The "super" for scientists is "cybernetic." For the philosophers, what is added is "telos." For the theologians, what is added to matter-energy is "logos." Genes do not contain simply descriptive information "about" but prescriptive information "for" the vital processes of life. There is natural selection "for" what a gene does contributing to adaptive fit. Stored in their coding, genes have a "telos," an "end." Magmas crystallizing into rocks and rivers flowing downhill have results but no such end. Genes are *teleosemantic*.

That differentiates physics from biology, and, biologists argue that they need to be alert to this. George C. Williams is explicit: "Evolutionary biologists have failed to realize that they work with two more or less incommensurable domains: that of information and that of matter. . . . The gene is a package of information" (quoted in Brockman 1995:43). In living things, concludes Manfred Eigen, this is "the key-word that represents the phenomenon of complexity: information. . . . Life is a dynamic state of matter organized by information" (1992:12, 15). John Maynard Smith says: "Heredity is about the transmission, not of matter or energy, but of information" (Maynard Smith 1995:28).

James A. Shapiro concludes: "Thus, just as the genome has come to be seen as a highly sophisticated information storage system, its evolution has become a matter of highly sophisticated information processing" (Shapiro 2002:10; 2005). The genome, a reservoir of previously discovered genetic know-how, is both conserving this and constantly generating further variations (new alleles), tested in the life of the organism (the phenotype). The better adapted (better informed) variants produce more descendants.

What is novel on Earth is this explosive power to generate vital information. In this sense, biology radically transcends physics and chemistry. It is not just the atomic or astronomical physics, found universally, but the middle-range earthen system, found rarely, that is so remarkable in its zest for complexity. Massive amounts of information are coded in DNA, a sort of linguistic or cognitive molecule. Now the semantic content is critical, as it was not in the minimal, mathematical, physical sense of information.

Genes are sometimes executive, as with the assembly of an embryo. But when the organism has been constructed and is launched into ongoing metabolism, the phenotypic organism becomes equally executive. The organism uses its genes as a sort of Lego kit where it finds the assembler codes for the materials it comes to need. Such complexity involves emergence. The mutual interactions of the components and subsystems results in a capacity and behavior of the whole that transcend and are different from those in the parts and unknown in the previous levels of organization. This proactive networking of genes and organism in environment involves a top-down causation (Noble 2006:42–54). There are hierarchal levels of control and the behavior of the phenotype (hungry and hunting) governs the metabolism (adrenalin levels or glucose use). The stored genetic information makes possible layered modular structures with complex functionality sensitive to environmental context.

The world of matter-energy, inherited from the first big bang, when taken over by the world of life at the second big bang, proves to be plastic enough for an organism to work its program on—as well as, at the third big bang, for a mind to work its will on. Although we cannot imagine any life without matter and energy, an organism has achieved the power to constitute the conditions under which appear the material energetic phenomena with which it interacts. Within this interaction, it can coagulate affairs this way and not that way, in accordance with its cellular and genetic programs. The macromolecular system of the living cell is influencing by its interaction patterns the behavior of the atomic systems.

Physicists find that a laboratory apparatus that humans have fabricated can constitute the conditions under which some phenomena appear, and within those conditions, can further coagulate these and not those specific phenomena from among the superposed quantum states. The actual phenomena that come to pass are interaction phenomena, if sometimes also, in other ways, random phenomena. Something similar is going on in organisms, but it is much more sophisticated than in the relatively crude physicist's machinery, which converts the atomic events into a photographic trace or a Geiger counter click.

The organism converts the phenomena into life. This is taking place with instrumental control much closer to the atomic level in a pervasive, systematically integrated way in the organism, while in the bulky physicist's apparatus we can manipulate processes and fabricate the materials of our instruments directly at the gross macroscopic levels, and only very indirectly at the molecular levels. But the organism is fine-tuned at the molecular level to nurse its way through the quantum states by electron transport, proton pumping, selective ion permeability, DNA encoding, and the like. Catalysis is especially impressive; of the thousands of metabolic reactions, virtually none would occur without an appropriate enzyme—mostly complex proteins. The organism via its information and biochemistries participates in forming the course of the microevents that constitute its passage through the world. The organism is responsible, in part, for the microevents, and not the other way around.

The organism has to flow through the quantum states, but it selects the quantum states that achieve for it an informed flow-through. The information within the organism enables it to act as a preference sieve through the quantum states, by interaction sometimes causing quantum events, sometimes catching individual chance events that serve its program, and thereby the organism maintains its life course. The organism is a whole that is program laden, that executes its lifestyle in dependence on this looseness in its parts. There is a kind of "downward causation" that complements an upward causation (Campbell 1974), and both feed on the openness, if also the order, in the atomic substructures. The microscopic indeterminism provides a looseness through which the organism can steer itself by taking advantage of the fluctuations at the micro levels. Life makes matter count. It loads the dice. Biological events are superintending physical ones. The organism is "telling nature where to go." Biological nature takes advantage of physical nature. The discovery that information is a critical determinant of organic-evolutionary history has thrown the causal/contingency debate into a new light.

Some leading theoretical biologists are now calling this genetic information "intentional," using that word in a nonconscious sense. John Maynard Smith claims: "In biology, the use of informational terms implies intentionality" (Maynard Smith 2000:177). That word has too much of a "deliberative" component for most users, but what is intended by "intentional" is the directed life process, going back to the Latin: *intendo*, with the sense of "stretch toward" or "aim at." Genes have both descriptive and prescriptive "aboutness"; they stretch toward what they are about. Kim Sterelny and Paul E. Griffiths speak of "intentional information" in contrast to "causal information." "Intentional information seems like a better candidate for the sense in which genes carry developmental information and nothing else does" (Sterelny and Griffiths 1999:104).

Intentional or semantic information is for the purpose of ("about") producing a functional unit that does not yet exist. Here there arises the possibility of mistakes, of error, and genes have some machinery for "error correction." None of these ideas makes any sense in chemistry or physics, geology or meteorology. Atoms, crystals, rocks, weather fronts do not "intend" anything and therefore cannot "err." A mere "cause" is pushy but not forward looking. A developing crystal has the form, shape, location it has because of, on the cause of, preceding factors. A genetic code is a "code for" something, set for "control" of the upcoming molecules that it will participate in forming. There is proactive "intention" about the future. This line of analysis confirms the actively cybernetic nature of biology.

Explosions: Combinatorial and Evolutionary

With the coming of life with such informational capacities, there appears another kind of explosion: combinatorial explosion. The numbers in astronomy are huge, on the order of 10^{80}, the number of protons in the visible universe (Barrow 2002:97–118). But when amino acids are linked together to construct proteins, the possible structures are immense (often defined as greater than 10^{110}) for proteins of about 100 amino acids; most proteins are over twice that long. "Because there are 20 different amino acids and a typical protein comprises some 200 of them, the number of possible proteins is greater than 20^{200} [or about 10^{260}]. . . . All of the matter in the myriad galaxies of the universe falls far short of that required to construct but one example of each possible protein molecule" (Scott 2002:297).

Living organisms can sometimes construct proteins quite fast, in fractions of a second. But the universe from big bang to present would have to be repeated 10^{67} times to create each one of these possible proteins just once. Typical DNA strands in mammals, with some hundred million base pairs, can be arranged in 10 raised to the 10^8 power different ways (Scott 1995:29–30). There is an explosion of possibilities in complication. Such vast possibility space is a mathematical computation; many of these theoretical possibilities are not in actual, empirical possibility space in the living world, owing to various constraints in construction and function. Nevertheless the possibilities are immense (Noble 2006:23–32). Consider in analogy the number of sentences that can be typed on a keyboard with 26 alphabetical letters, upper and lower case, some punctuation marks, spaces, numerals, a number in the range of 100 keys. Nature on Earth rings the changes on these biomolecular possibilities, exploring biodiversity in adaptive fit.

Astronomical nature and atomic nature, profound as they are, are nature in the simple. At both ends of the spectrum of size, nature lacks the complexity that it demonstrates at the mesolevels, found in the earthen ecosystem, or at the psychological level in the human person. Astronomical nature is incredibly vast and energetic, but primitive. Such a statement will seem odd, on first reading, for the theories and calculations by which the mind probes such nature are among the most sophisticated known to science—for example, relativity theory and quantum mechanics. Physics is no simple science, and the stuff of its observations is abundantly mysterious, as the considerations we undertook in the previous chapter should reinforce. But that energetic matter, compared with life and mind, is as primitive as it is basic. We encounter advanced forms of natural organization only at the middle ranges and in the other sciences. We humans do not live at the range of the infinitely small, nor at that of the infinitely large, but we may well live at the range of the infinitely complex.

There is in a typical handful of humus, which may have 10 billion organisms in it, a richness of structure, a volume of information (trillions of "bits"), resulting from evolutionary processes across a billion years of history, greatly advanced over anything in myriads of galaxies, or even, so far as we know, all of them. Information accumulates across the spectrum of species, stored genetically. Further, there emerges, explosively with the development of neurons and brains, powers of acquired learning. These combine in an especially impressive way in humans. The human being starts out as a single cell and the information in that genome generates with increasing complexity a highly functional organized body with 10^{13} to 10^{14} cells of more than 200 cell types. If the DNA in the myriad cells of the human body were uncoiled and stretched out end to end, that microscopically slender thread would reach to the sun and back over half a dozen times. The human being is the most sophisticated of known natural products. The human brain, built by DNA, is the most complex entity known in the universe, a consideration we will further develop with the third big bang.

You may reply that this took several billion years, so thinking of life on Earth as a genetically based explosion is misplaced. The main idea in evolution is incremental, gradual unfolding bit by bit. Half of the history of life was as one-celled procaryotes, which may not seem explosive, although recent estimates find microbial biodiversity immense.

By one account, evolution is characterized by millions of years of stasis, punctuated by relatively brief periods of rapid change (Gould and Eldredge 1977). Evolution is seldom rapid; mostly it is quite boring. Nor is it always expanding. Mass extinctions can cause great loss of diversity.

The first big bang continues as an expanding universe, a 13 billion years' explosion. The second big bang has continued expanding life on Earth, a 3 billion years' explosion. Fast or slow is a matter of scale, both in space and on Earth, and at astronomical and atomic levels. If evolution takes centuries, the molecular biology that supports it can be fast. A single *Escherichia coli* cell, dividing every hour, synthesizes per second 4,000 molecules of lipid, almost 1,000 protein molecules, each containing about 300 amino acids, and 4 molecules of RNA. Some processes are autocatalytic; they feed back on themselves and escalate the speed of the process. Positive feedback can escalate and drive exponential changes. Allosteric enzymes can amplify metabolic processes a million times over the spontaneous rate without such enzymes. Other enzymes can accelerate reactions up to 10^{20} times the uncatalyzed reaction (Ringe and Petsko 2008). A developing fetal brain can generate over a quarter million neurons in the minute during which the reader is here pausing to consider whether the life processes are fast or slow (Cowan 1979). DNA replication, if it were expanded to the everyday world, would take place at the speed of a jet plane.

Yes, the evolution of life, like the evolution of stars, takes time, lots of it, especially to reach higher forms of life. In this evolution natural systems were often sustained in the past for long periods, even while they gradually modified. Some of the feedback loops, the feed loops, will seek equilibrium states, suppressing dynamic change. Negative feedback can stabilize processes, just as positive feedback can amplify them. Natural selection means continual changes, but it fails without order, without enough stability in ecosystems to make the mutations selected for dependably good for the time being.

There is genetic information, but such information must be appropriate to and operative in a stimulating environment, with feedback and feed-forward loops. Spring weather wakes up seeds, dry weather forces deeper roots, and both sorts of weather turn on genes to sprout, to root. Both weathers select for plants with the capacity to do this well. There is variation, more or less contingent, but without relative stability in environments, sustained patterns of evolutionary change cannot occur.

A rabbit with a lucky genetic mutation that enables it to run a little faster has no survival advantage to be selected for, unless there are foxes reliably present to remove the slower rabbits. Ecosystems have to be more or less integrated (in their food pyramids, for example), relatively stable (with more or less dependable food supplies, grass growing again each spring for the rabbits), and with persistent patterns (the hydrologic cycle watering the grass), or nothing can be an adapted fit, nor can adaptations evolve.

Evolution cannot occur in a world too chaotic to provide reliable life support, nor in one too stable to provide challenging new opportunities. The system is, from the short-term perspective, in equilibrium, even when from a longer-range perspective it is exploding. The explosion feeds off the equilibrium. Relatively simple individuals can form complex associations, ecologies, and this ecological complexity provides new niches that challenge the development of new forms of life. This is called "bootstrapping in ecosystems," feed-forward loops that generate new niches that reinforce each other and open up new opportunities for species specialization (Perry et al. 1989). Cumulatively over the millennia, owing to the genetic capacity to acquire, store, and transmit new information, complexity can increase. There are advantages in specialized cells or organs, the efficiencies of the division of labor, and this couples more complex and more diverse forms of life.

A diverse environment is heterogeneous, and species are favored that are multiadaptable, not just well adapted to one homogeneous environment. Such adaptability requires complexity, capacities to search out better environments and migrate to them, and, once there, capacities to invade successfully, to prey on or resist predation by, or to find and share resources with, the different kinds of organisms that can live in both wet and dry, cold and hot, grassland and forested environments. Becoming more complex sometimes helps in dealing with the challenges and opportunities offered by diversity. Complexity helps in tracking changing environments.

Reptiles can survive in a broader spectrum of humidity conditions than can amphibians, mammals in a broader spectrum of temperature than can reptiles. Once there was no smelling, swimming, hiding, defending a territory, gambling, making mistakes, or outsmarting a competitor. Once there were no eggs hatching, no mothers nursing young. Once there was no instinct, no conditioned learning. Once there was no pleasure, no pain. But all these capacities got discovered by the genes. Once there was no metameric segmentation, as in worms; once there was no pentameric segmentation, as in starfish. But all these phenomena appeared, gradually, but eventually elaborating and escalating the diversity and complexity of life.

On geological scales, biologists can speak of explosions, as with the "Cambrian explosion," when complex multicellular life vastly elaborated (Valentine 2004:chapter 14; Knoll 2003:6). Since the Cambrian, Motoo Kimura estimates that higher organisms have accumulated genetic information continually to the present at the average rate of 0.29 bit per generation (Kimura 1961). Interestingly, mass extinctions, though initially setbacks, can trigger subsequent explosions. There is typically a fast rebound, but these are not simply replacements. Respeciation takes off in new directions. Douglas H. Erwin says: "The end-Permian mass extinction triggered an explosion in marine diversity and a reorganization of marine communities. . . . Similar changes occurred on land" (Erwin 1993:261). In the Cenozoic era, Philip W. Signor documents for both plants and animals "a spectacular diversification over the past 100 million years" (Signor 1990:529). The challenges of survival and relentless pressures for better-adapted fit have driven escalating skills from the Cambrian to the present.

Among paleontologists, there is continuing debate about how radical or rapid such transitions were: the relationship, for example, between the novel Cambrian fauna and the Ediacaran (pre-Cambrian) animals they replaced; whether at that time the plants evolved as rapidly as the animals (Vickers-Rich and Komarower 2007). In addition to the famous Burgess Shale in Canada, two more recently discovered Cambrian sites, Chengjiang in China and the Sirius Passet formation in Greenland, show a rich and diverse fauna, supporting the claim that the Cambrian radiation was truly explosive. Always, on geological scales, fast or slow is a matter of perspective; still, two dozen paleontologists analyze what they term "major evolutionary radiations" across evolutionary history. "Short-lived, highly 'creative' evolutionary phases of rapid diversity increase and morphological divergence characterize the evolutionary history of many groups" (Taylor and Larwood 1990:vi). There are repeated spurts of evolutionary novelty and resulting adaptive radiation.

Paleontologists also recognize critical thresholds, revolutions in the evolutions, so to speak. John Maynard Smith, the dean of theoretical biologists, says: "Heredity is about the transmission, not of matter or energy, but of information. . . . The concept of information is central both to genetics and evolution theory" (Maynard Smith 1995:28). He and his colleague, Eörs Szathmáry, analyze "the major transitions in evolution" with the resulting complexity, asking "how and why this complexity has increased in the course of evolution." "Our thesis is that the increase has depended on a small number of major transitions in the way in which genetic information is transmitted between generations" (Maynard Smith and Szathmáry 1995:3).

Maynard Smith finds that each of these innovative breakthrough levels is surprising, not scientifically predictable on the basis of the biological precedents, but with dramatic results. What makes the critical difference in evolutionary history is increase in the information possibility space, which is not something inherent in the precursor chemical and physical materials, nor in the incrementally developing evolutionary system. The vital genesis of emergent information channels punctuates evolutionary history. This happens when the genetic code is discovered, when DNA becomes concentrated in the nucleus, with the appearance of meiotic sex, transforming capacities for information transfer. Another threshold is crossed with multicellular life, with specialized capacities for organ hierarchies (livers, kidneys, legs), elaborating large networks for distributed discrete functions. Yet another transformation comes with the generation of neurons and brains, with their capacities for acquired learning, for language. Each transformation breaks former constraints and escalates evolutionary possibilities.

Genes speciate new kinds in response to environmental challenge, as much as they reproduce existing kinds to the maximum possible extent. Maynard Smith pinpoints the limitless potential novelty in genetics: "There are today, in the living world, only two systems capable of unlimited heredity, that is, of transmitting an indefinitely large number of different messages: these are the genetic system based on nucleic acids, and human language" (Maynard Smith 2000:215). That couples, interestingly, what we are here calling the second and the third big bangs. "Descent with modification" results in ever-ongoing "ascent with modification." As Francisco Ayala puts it, there is "increase in the ability to gather and process information about the environment" (Ayala 1974:344). Genes are set to generate biodiversity without end.

A genetic sequence has a potential for being an ancestor in an indefinitely long line of descendant genotype/phenotype reincarnations. Genes act directed toward a future, under construction, producing a functional unit that does not yet exist. In physics and chemistry as such, there can be only sources, never resources. In biology, the novel resourcefulness lies in the epistemic content conserved, developed, and thrown forward to make biological resources out of the physicochemical sources. A crucial line, a singularity, is crossed when abiotic formations get transformed into loci of information. The *factors* come to include *actors* that exploit their environment.

A genome conserves a form of life, but a genome is equally a search program. Genes are as dynamic as they are stable units. Mutation, crossing over, drift, allelic variation, cutting and splicing, insertions, deletions—all this disrupts conserving the inherited genes; but such processes also make genes information generators. There is "facilitated variation" (Kirschner and Gerhart 2005). Genetic sets may be tested for their "evolvability," their flexibility in encounters with shifting environments (Kirschner and Gerhart 1998). Genes generate trial-and-error solutions, some of which yield novel information discovered. Genes interweave possibilities and explore new possibility space. A result of this "highly sophisticated information processing" (recalling Shapiro 2002:10; 2005) is open-ended, explosive radiation of biodiversity. Genes in living systems explore a combinatorially immense space of possibilities through the evolutionary process.

The construction of complexity begins with the biologically simple, one-celled organisms; the story is from the bottom up. After certain thresholds are reached, parts can be differentiated. There is division of labor, a principal result of multicellularity. Now evolutionary development compounds the growth of biodiversity and biocomplexity. Often these relations are nonlinear; causes and effects are not proportionally related. Organic systems build by reiteration and modification of modules (mutations enabling survival in colder weather), though there may also be novel sorts of modules (enabling in mammals nursing mothers, unknown in reptiles).

These biological achievements are quite impressive, wonderful. But now the generative context of order and contingency no longer seems fine-tuned. Rather it is often groping, exploratory, with false starts and setbacks, ragged and wasteful. Indeed, at the same time that astrophysics and nuclear physics have been describing a universe "fine-tuned" for life, evolutionary and molecular biology seem to be discovering that the history of life is a random walk with much struggle and chance, driven by selfish genes. The process is prolific, but no longer fine-tuned. To the contrary, evolutionary history can seem tinkering and makeshift at the same time that, within structural constraints and mutations available, it optimizes adapted fit. Mutations are random, unrelated to the needs of the organism; most are worthless or harmful. Mutations may have causes (a cosmic ray, a chemical mutagen absorbed), but these involve the intersection of unrelated causal lines, statistical probabilities, or even (with a mutation caused by radioactive decay) quantum indeterminism—the latter a sort of causal gap.

Even if, more comprehensively seen, the genetic searching is highly sophisticated information processing, it involves trial and error, in which mutations play a part. When immense possibility spaces open up in the combinatorial explosion, fine-tuning no longer seems the relevant, the desirable, or even the possible route of creativity. The heavens may be clockwork, but the evolutionary epic is adventure. Creativity is more open-ended, more opportunistic in an ambiguous and challenging world. On Earth there is wild nature, wilderness, and there is something almost logically contradictory about fine-tuning the wild. Fine-tuning may be required to produce the elements of life, but the struggle for adapted fit is at another level of creativity. Recalling Andrew Knoll, "life's long history abounds in . . . narrative verve" (Knoll 2003:xi). That struggle, ongoing in wild nature, can produce some quite well-adapted fits, impressively competent at both phenotypic and genotypic levels. If you wish to phrase it this way, the evolutionary epic has been a struggle for ever better "tuning" of life in its environments. The results can be "fine-tuned," even if the process cannot.

Biodiversity and Biocomplexity

There are two dimensions to the explosions involved in the second big bang: biodiversity and biocomplexity. Life started in the seas, so we should first look at the marine record. J. W. Valentine, after a long survey of evolutionary history, concludes for marine environments that both complexity and diversity increase through time. First, with regard to diversity:

> A major Phanerozoic trend among the invertebrate biota of the world's shelf and epicontinental seas has been towards more and more numerous units at all levels of the ecological hierarchy. This has been achieved partly by the progressive partitioning of ecospace into smaller functional regions, and partly by the invasion of previously unoccupied biospace. At the same time, the expansion and contraction of available environments has controlled strong but secondary trends of diversity. . . . The biosphere has become a splitter's paradise. (Valentine 1969:706)

There are ups and downs in numbers of families and species, due to the contingencies of drift; nevertheless, biologically, the trend is upward.

Complexity also increases:

> A sort of moving picture of the biological world with its selective processes that favor increasing fitness and that lead to "biological improvement" is projected upon an environmental background that itself fluctuates. . . . The resulting ecological images expand and contract, but, when measured at some standardized configuration, have

a gradually rising average complexity and exhibit a gradually expanding ecospace. (Valentine 1973:471)

Environments fluctuate, as Valentine recognizes. We worried with the primordial explosion of matter-energy that explosions can be messy, and found that there is a resulting mixture of both order and disorder. The explosion of life on Earth is cybernetic, but now there are crooked lines, ups and downs, twists and turns, as one might expect when billions of species evolve over billions of years. The history of life is something of a thicket. Just how explosive which parts of marine invertebrate evolution were remains under analysis, with some recent studies agreeing that the rise was sometimes steep but also less steep than previously thought in more recent times (Alroy et al. 2008). Others still find "dramatic evolutionary radiations" (Stanley 2007:11). Much of the debate turns on the extent of biases in the fossil record, sampling errors, regional collecting, local versus global trends, ice ages.

A common interpretation of the periods of stasis in the seas is that Earth's tectonic plates were configured to fuse the landmasses, resulting in a saturation of kinds of species that had at that point evolved on the continental shelves. Since marine life is primarily on the continental shelves, it may be especially susceptible to contingencies in continental drift. Meanwhile, there seems little doubt that in the seas there are both fluctuations and gradually rising diversity, which can sometimes be exponential.

M. J. Benton summarizes the global picture in three graphs (figs. 2.1–3, Benton 1995), which convey several impressions. One is that the number of families starts low and ends high, alike for all organisms on land and in sea. Another is that life's expansion speeds up in later geological time, although the diversifying processes of early life may have fossilized poorly or be lost in the ancient fossil record. In marine life there is a flat part of the curve from Ordovician to Permian times. But one has also to remember that in that period life moved from sea to land, which demanded considerable novel complexity in speciation, since the terrestrial environment is more demanding. By species count, most of the world's plants and animals live on land, despite the much larger sea environment. Life in the sea does retain more basic animal

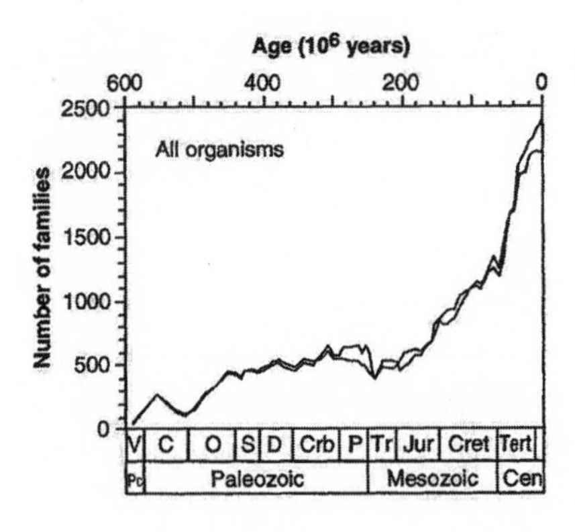

Figure 2.1 Diversification of life through time for all organisms (*Benton 1995*)

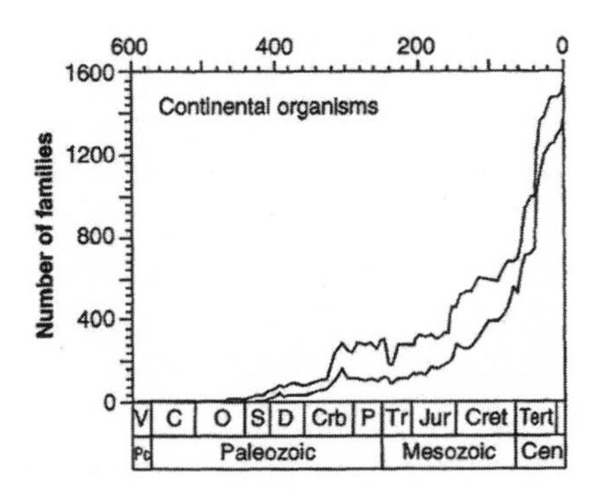

Figure 2.2 Diversification of life through time for continental organisms (*Benton 1995*)

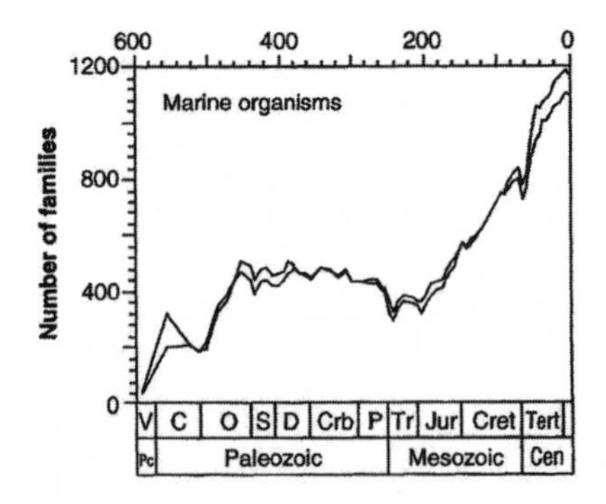

Figure 2.3 Diversification of life through time for marine organisms (*Benton 1995*)

forms (thirty-five phyla) than life on land (ten phyla), but the overall species count on land is some ten times higher than in the sea.

Plants develop steadily on the land masses, with species turnover resulting in increased diversity. Andrew H. Knoll graphs (fig. 2.4) this record for local ecosystems over evolutionary time. In the Paleozoic there is a general rise, and after that a plateau. "The history of diversity within floras from subtropical to tropical mesic floodplains is marked by several periods of rapid increase separated by extended periods of more or less unchanging taxonomic richness." After the mid-Mesozoic, with the rise of the angiosperms (flowering plants), there is a steady climb in regional floras. Knoll concludes "that species numbers within subtropical to tropical communities have been rising continually since the Cretaceous and that a plateau has yet to be established" (Knoll 1986:140, 132; Wilson 1992:195).

In the composition of the floras and faunas, certain forms can later be less numerous than before; but, climatic conditions permitting, overall biodiversity gradually and sometimes rather spectacularly rises, as seen in graphs illustrating the composition of vascular plants (fig. 2.5, Niklas 1986) and vertebrate orders (fig. 2.6; Niklas 1986). Here

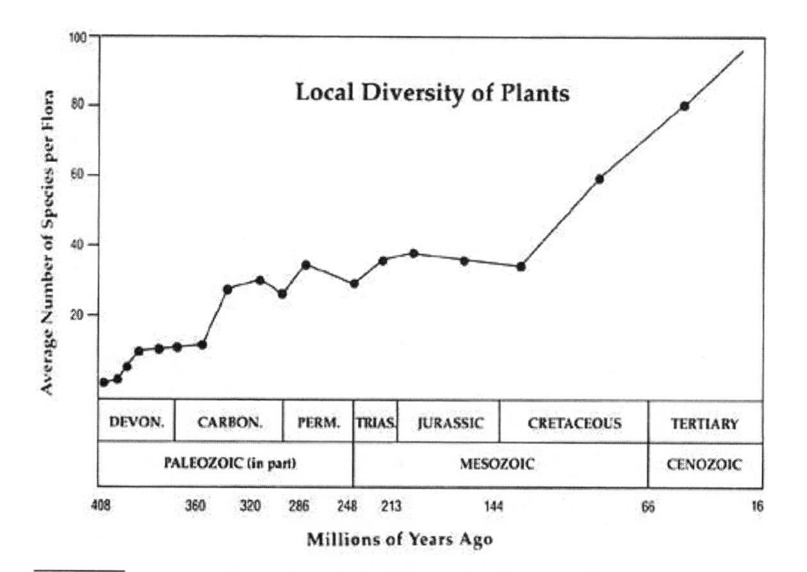

Figure 2.4 Average number of plant species found in local floras (*Wilson 1992, adapted from Knoll 1986*)

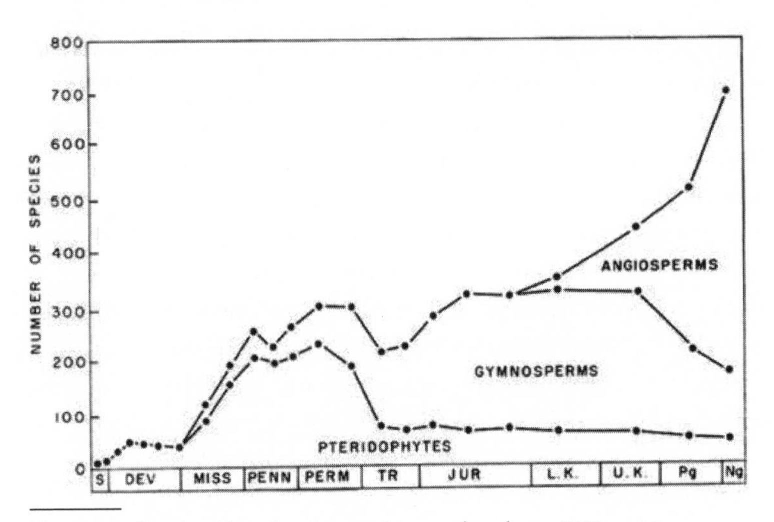

Figure 2.5 Species diversity changes in vascular plants (*Niklas 1986*)

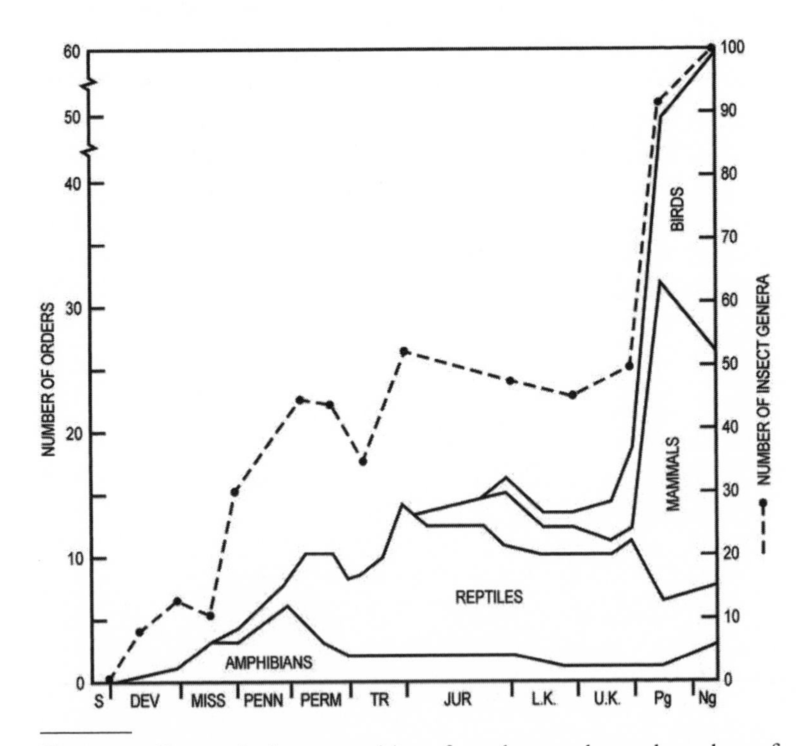

Figure 2.6 Changes in the composition of vertebrate orders and numbers of insect genera (*Niklas 1986*)

too the later-coming forms are often more complex than the earlier ones they replace. Mammals with their warm blood and higher energy requirements develop metabolisms and behavioral skills not present in cold-blooded reptiles and amphibians. Angiosperms advance over, and may displace, gymnosperms. Fortunately for overall biodiversity, these earlier groups, in reduced numbers (and with species turnover), often continue to enrich present faunas and floras.

Consider a series of graphs in land plant diversification (Niklas, Tiffney, and Knoll 1985:193, 107, 112). The first graph (fig. 2.7), Silurian through Devonian times, escalates up, then down. The second (fig. 2.8), Carboniferous to Lower Cretaceous, shows escalation, then stagnation in total numbers, but with erratic shifts in the kinds of plants—ferns, lycopods, cycads, conifers. The third (fig 2.9), Lower Cretaceous

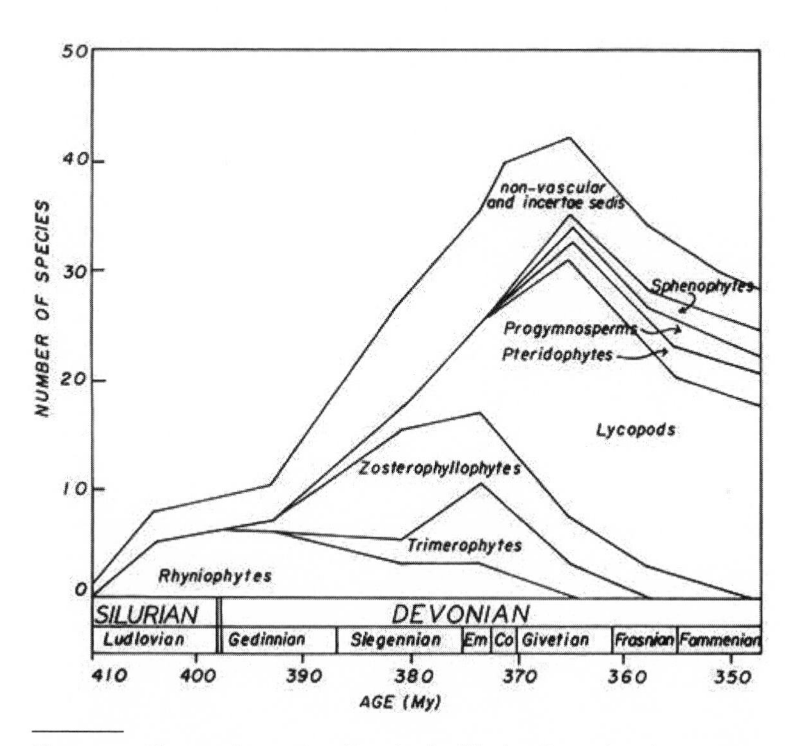

Figure 2.7 Changes in species diversity in Silurian-Devonian times *(Niklas, Tiffney, and Knoll 1985)*

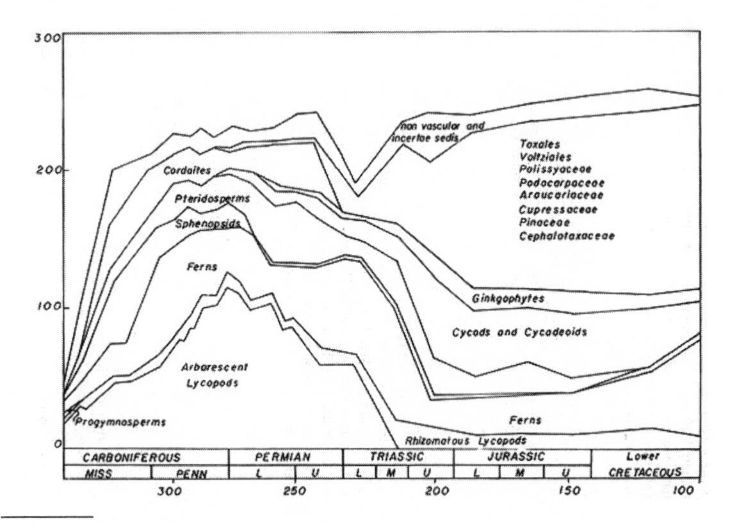

Figure 2.8 Changes in species diversity in Carboniferous-Lower Cretaceous times (*Niklas, Tiffney, and Knoll 1985*)

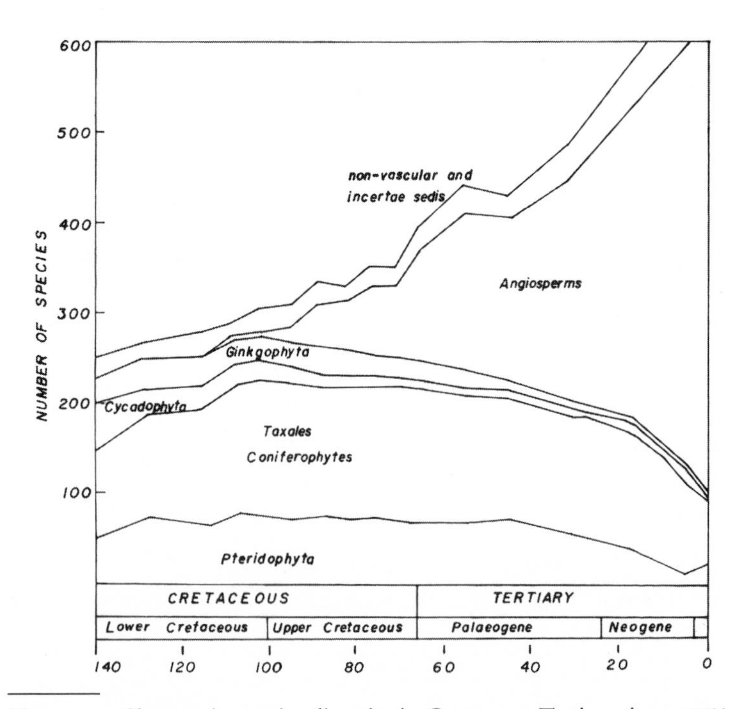

Figure 2.9 Changes in species diversity in Cretaceous-Tertiary times (*Niklas, Tiffney, and Knoll 1985*)

to recent times, shows steady rise, dominated by angiosperms, and older forms declining.

But if we watch the scales of the graphs and align them for an overall record, the result is a rather striking graph of exploding plant life, with overall increases in biodiversity, as much because of as despite fluctuations in kinds (fig. 2.10). (Subgroups are placed differently in the three graphs.)

Turning to animals, consider a graph of increasing diversity of orders in vertebrates (fig. 2.11, Padian and Clemens 1985:50). Again the graph, in the first half, has a jagged look, but the overall impression is strikingly of increasing vertebrate diversity over geological time.

E. C. Pielou concludes a long study of diversity: "Thus worldwide faunal diversification has increased since life first appeared in a somewhat stepwise fashion, through the development and exploitation of adaptations permitting a succession of new modes of life" (Pielou 1975:149). Life appears in the seas, moves onto the land, then into the skies. Terrestrial communities developed from the Silurian onward. In the Tertiary there was a marked increase in diversity due to the rise of warm-blooded vertebrates (mammals and birds), more than making up for the decrease in reptiles and amphibians. When vertebrates took to the air, there was introduced an entirely new mode of life.

There were setbacks, notably in the Permian-Triassic and again in the wave of mammal extinctions in the middle (pre-Pleistocene) Quaternary. But there was recovery. Many factors figure in, including climates and continental drift. Sometimes, the change due to organic evolution is overwhelmed by the change due to climatic cooling or drying out. The change due to organic evolution may be accelerated or decelerated by continental drift; continents fused together may provide a bigger area that supports more species, or they may provide more competition that eliminates species that previously evolved on separated continents. If the tectonic plates drift together and form a supercontinent, the supercontinent may saturate (some think), and if afterward the continents drift apart, this may add to the provinciality of the world and facilitate by isolation the evolution of diversity. On the whole organic evolution has "the result that the present diversity of the world's plants and animals is (or was just

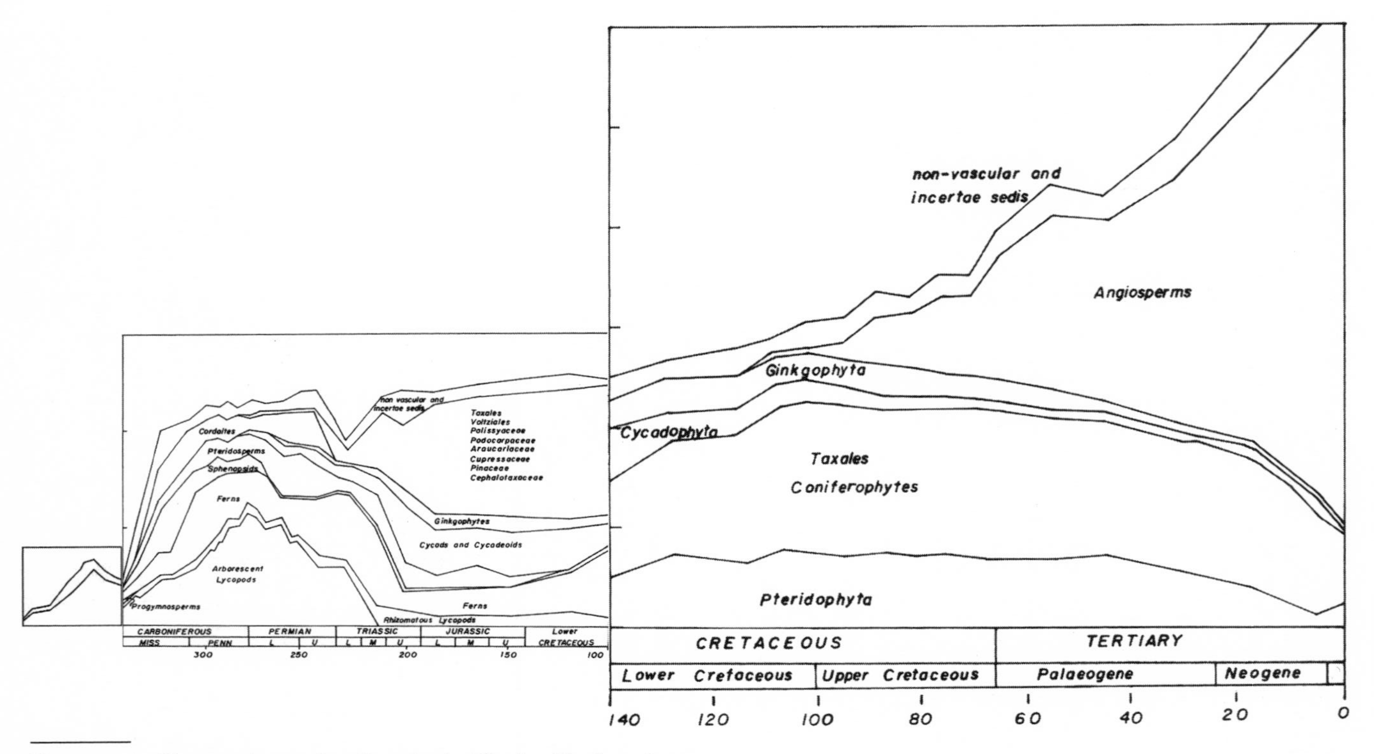

Figure 2.10 Changes in species diversity in Silurian-Tertiary times

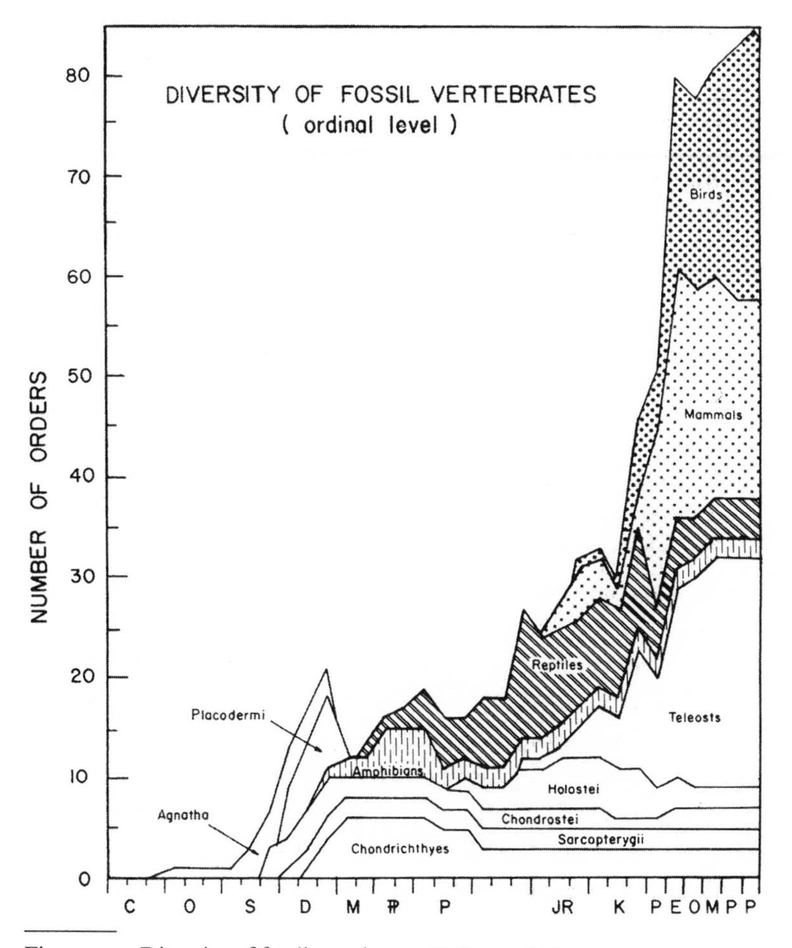

Figure 2.11 Diversity of fossil vertebrates *(Padian and Clemens 1985)*

before our species appeared) probably greater than it has ever been before" (Pielou 1975:150).

John Bonner, in a detailed study of the evolution of complexity, summarizes his findings:

> There has . . . been an extension of the upper limit of complexity during the course of evolution. . . . There has also been an increase in the complexity of animal and plant communities, that is, there

> has been an increase in the number of species over geological time, and this has meant an increase in species diversity in any one community. . . . One can conclude that evolution usually progresses by increases in complexity. . . . As evolution proceeded on the surface of the earth, there has been a progressive increase in size and complexity. (Bonner 1988:220, 228, 245)

Life's big bang explodes in an extravaganza, in both diversity and complexity.

Expanding Neuro-sentience: Felt Experience

With the emergence of neurons, arranged in neural networks, this exploding complexity reached sentience. This continued the life explosion, deepening and transcending the cybernetic dimensions. A neuron is functionally "for" information detection and transfer. Neurons make possible acquired learning, acquiring information and storing it for future use in the lifetime of the individual. Behavior is more labile, less stereotyped. With neuronal nets of increasing complexity, a new threshold is crossed; there appears felt experience. As with everything else in evolutionary development, this takes place gradually, but that ought not to obscure the fact that there is momentous emergent novelty. With increasing neuronal complexity, there appears *inwardness*. With still more, there appear what philosophers call *qualia*, consciously entertained experiential mental states such as sensations, feelings, perceptions, desires. Increasingly, there is "somebody there"; objects evolve into subjects with psyche. There is "something it is like" to be that subject-organism. There appears phenomenology of experience, as when a person (or a rat?) smells strong cheese.

Analyzed at the molecular level, neurons involve the flow of various ions, electric currents, often across membranes, as with synaptic connections. These processes can as well be considered complex chemistries. Living organisms, from the start of life, have been doing non-nervous metabolisms interacting with ions and their bonding chemistries, including some signaling and stimulus-response perceiving. Plants detect and act upon signals within their environments. A protozoan may move up a gradient toward light or food. Sometimes biologists and philosophers have wondered whether low-grade felt awareness might accompany such perceiving even in non-neuronal organisms (Bray 2009; Jennings 1906:336–337), but most hold that felt experience, psyche, requires neurons.

Just when and how what might be a precursor to neurons appeared is not known, although there is some evidence this was about 700 million years ago. The diversity of existing nervous systems is enormous. Some scientists have wondered if nerves evolved independently more than once, although recent opinion, based on genetic and molecular analysis, indicates a single (monophyletic) origin (Hirth and Reichart 2007). The most primitive organisms to possess a nervous system are cnidarians, a phylum of mostly marine animals. In their diffuse nervous systems (as found in jellyfish, sea anemones, corals), nerve cells are distributed throughout the organism, often organized into nerve nets with synapselike connections, perhaps with ganglia, local concentrations of neurons. Sensory neurons connect with effector neurons without central integration. Presumably there is present some diffuse experience of feeling; it is difficult to know.

Central nervous systems evolved later. It is not known when they first appeared in animal evolution nor what their earliest function was. The presumed earliest ancestors are identified as "urbilateria," of which there are fossil traces (Arendt et al. 2008). Flatworms exhibit bilateral symmetry, breaking previously radial symmetry. This more is different. The nervous system evolved to consist of longitudinal nerve cords, with peripheral nerves connecting to sensory cells, and at one end a "brain," as for instance in the two joined cephalic ganglia in *Planaria*. There does appear to be present felt experience, though such mental states are simple, without capacity for introspection (Tye 1997). There are endorphins (natural opiates) in earthworms, which indicates both that they suffer and that they are naturally provided with pain buffers (Alumets et al. 1979).

Alain Ghysen finds that central nervous systems probably originated in a single species with a sophisticated enough nervous system to be both conserved in basic features and dramatically elaborated in the myriads of brained species that have characterized evolutionary history. "The appearance of very different life forms, lifestyles and habitats requires the previous attainment of a neural circuitry that is sufficiently robust to cope with large changes without losing its overall coherence" (Ghysen 2003:555). So nerve cells appear, and radically elaborate and escalate capacities across evolutionary history. There is a sense, however, in which calling this elaboration and escalation obscures the radical momentous innovation of subjectivity appearing where before was only objectivity.

It may seem that the evolutionary account is delivering subjectivity and felt experience bit by bit, incrementally rather than swiftly, but it is also true that felt experience appears where absolutely none was before. Incremental qualities joined and rejoined are also reformed and transformed into novel qualities. One gets, at length, pleasure and pain by organizing millions of unfeeling atoms. Slowing things down and putting together molecular parts does not really alleviate the lack of theory explaining how inwardness comes out of outwardness. It only spreads the inexplicable element thin, rather than asking us to swallow it in one lump.

We do not know how, much less why, there emerges out of the neural electronics and chemistry this capacity for conscious experience and caring. The molecular accounts of currents and chemistries in neurons describe the technical conditions necessary for the production of subjective experience, with no account of the necessary or intelligible derivation of what emerges. "Nobody has the slightest idea how anything material could be conscious. Nobody even knows what it would be like to have the slightest idea about how anything material could be conscious" (Fodor 1992:5).

The physical world that resulted from the first big bang could not feel pain or pleasure. But with advancing formational and informational levels, life crossed another singularity: increasing capacity for felt experience. A planet moves through an environment, but only an organism can need its world, a feature simultaneously of its prolife program and of the requirement that it overtake materials and energy. But if the environment can be a good to it, that brings also the possibility of deprivation as a harm. To be alive is to have problems. Things can go wrong just because they can also go right. In an open, developmental, ecological system, no other way is possible. All this first takes place at insentient levels, where there is bodily duress, as when a plant needs water.

Irritability is universally present in life; suffering in some sense seems copresent with neural structures. Sentience brings the capacity to move about deliberately in the world, and also to get hurt by it. We might have sense organs—sight or hearing—without any capacity to be pained by them. But sentience was not invented to permit mere observation of the world. It rather evolved to awaken some concern for it. Sentience coevolved with a capacity to separate the helps from the hurts in the world. A neural animal can love something in its world and is free to seek this, a capacity greatly advanced over anything known in immobile, insentient plants. It has the power to move through and experientially to evaluate the environment. The appearance of sentience is the appearance of caring, when the organism is united with or torn from its loves. The earthen story is not merely of goings-on, but of "going concerns." The step up that brings more drama brings suffering.

Further, pain is an energizing force. Suffering not only goes back-to-back with caring sentience but also drives life toward pleasurable fulfillment. The good presupposes concomitant evil, but the evil is enlisted in the service of the good. Individually, the organism seeks to be rid of pain, and yet pain's threat is self-organizing. It forces alarm, action, rest, withdrawal. It immobilizes for healing. The organism is quickened to its needs. The body can better defend itself by evolving a neural alarm system. The experiences of need, want, calamity, and fulfillment have driven the natural and cultural evolution of abilities to know, and in due course abilities to think. Where pain fits into evolutionary theory, it must have, on statistical average, high survival value, with this selected for, and with a selecting against counterproductive pain. In this sense, pain is a prolife force.

The evolutionary explosion is driven by conflict and resolution. All advances come in contexts of problem solving, with a central problem in sentient life the prospect of hurt. We do not really have available any coherent alternative models by which, in a painless world, anything like the explosion of life on Earth might have taken place. The system summarizes the lives of individuals in their conflict and resolution, using this to innovate by spinning out biodiversity and biocomplexity. Adapted fit seems at first a good thing, but then the shadow side is how each organism is doomed to eat or be eaten, to stake out what living it can competing with others. Perhaps there is more efficiency than waste, more fecundity than indifference, but each organism is ringed about with competitors and limits, forced to do or die. Each is as much set against the world as supported within it.

Seen more systemically, the context of creativity logically and empirically requires this context of conflict and resolution. An environment entirely hostile would slay life; life could never have appeared within it. An environment entirely irenic would stagnate life; advanced life, including human life, could never have appeared there either. Oppositional nature is the first half of the truth; the second is that none of life's explosive advance is possible without this dialectical stress. Muscles, teeth, eyes, ears, noses, fins, legs, wings, scales, hair, hands, neurons, brains—all these and almost everything else comes out of the need to make a way through a world that mixes environmental resistance with environmental conductance.

The system, from the perspective of the individual, is built on competition and premature death. Seen systemically, that is the generating and testing of selves by conflict and resolution, such values required to be both prolific and adapted fits. Such conflict in resolution does result in better-adapted fit, where the organism occupies a niche providing life support in an ecology of interdependent, mutually supporting species. Partnerships and symbioses too are vital in the evolutionary history of life.

The evolutionary story could be titled, "The Evolution of Caring," positively the capacity to enjoy pleasures. Or, perversely if one prefers, "The Evolution of Suffering." Each seeming advance—from plants to animals, from instinct to learning, from sentience to self-awareness, from nature to culture—steps up the pain. It is difficult to extrapolate to animal levels and make judgments about the extent of their suffering. A safe generalization is that pain becomes less intense as we go down the phylogenetic spectrum, and is often not as acute in the nonhuman as in the human world (Eisemann et al. 1984).

No doubt there was an evolutionary genesis of neurally based mind, capable of conscious pleasures and pains. But we have no logic by which out of physical premises, one derives biological conclusions, and taking these as premises in turn, one then derives psychological conclusions. Are neurons and the consciousness they produce somehow precontained in the first big bang? Are they even precontained in the second? Nevertheless, there is no doubt that animal life gets psyche, and that higher forms of animal life get psyched up. Matter starts to get animated, spirited. Maybe we are here getting hints of headings toward the third big bang?

Escalating Co-option: Serendipity

A major feature of genetic natural history is co-option generating novel, nonlinear possibilities. An existing gene and its product are recruited to a new function, with serendipity. For example, lens crystallins used in eyes first evolved in an altogether different role, as heat-stress proteins. They happened to be transparent. Surprisingly, they get used to make eye lenses, and more than once (Wistow 1993). Darwin had already noted that this could happen: "The swimbladder in fishes . . . shows us clearly the highly important fact that an organ originally constructed for one purpose, namely flotation, may be converted into one for a wholly different purpose, namely, respiration" (Darwin 1968:220–221). So the movement of life from sea to land was by co-option.

There are remarkable forks off preexisting pathways. Often, though not always, there is gene duplication and one copy serves the former function while the new copy can be modified in exploratory directions. Previously disconnected parts working along unrelated pathways are co-opted off and put together to start serving a novel function, perhaps only slightly well at first. Radically different selection pressures begin to work in new directions that are completely unanticipated when they occur. Once launched, the novel functions may improve steadily and completely transform the course of natural and human history.

Perhaps it all takes place by slight modifications of a precursor system. But these slight modifications are now being made in new, unprecedented directions. The co-opting modification is not improving the initial function but angles off in a new direction. The change is not iterative; it is metamorphic. Co-option breaks up channelized and entrenched developmental lines and opens up new directions. Restriction enzymes, one of the most important features of genetic innovation and a principal tool in genetic engineering, were first invented by bacteria to cut their parasites into pieces. They turned out to be useful for organisms to cut their own genomes into pieces and reshuffle them in the search for co-options.

Environmental and structural constraints remain but the constraints are not what they were before, now that the organism is equipped with these new potential capacities. The amount of information in an organism is transforming into its capacity for self re-formation, though the self-re-formation is also provoked, evoked by environmental challenge and stress. Self-organizing becomes self-transcending.

Co-option can escalate cybernetic capacities. Consider the evolution of hearing, vital for animals needing information about their world (Bear, Conners, and Paradiso 2001:chapter 11). Hearing evolved from cells in the side of an aquatic vertebrate's body that were sensitive to pressure, helpful to a swimming animal, an original use that has been lost from the reptiles onward. These cells were co-opted to become the hair cells in mammalian ears. That required constructing the external, middle, and inner ears, with small bones co-opted and modified to amplify sound, vibrating an oval area on the cochlea of the inner ear. This jiggles the microscopic hairs (stereocilia) on the ends of the hair cells. These cells synapse with neurons. The hairs are sensitive to movements as small as 0.3 nanometers (about the diameter of a large atom). Mechanical movement of the cilia opens and closes ion channels, letting sodium ions into the cell, and this constitutes an electric current, which triggers the synapsizing, producing perceptible noise, over a volume differential of a trillion times from softest to loudest.

Animals need to know frequencies as well as volume, and here the firing frequencies of the usual synaptic transmissions can track frequencies at the lower ranges, but the higher frequencies are too fast for this method. So ears improvise something else. There has further evolved a basilar membrane packed with hair cells and rolled up in the cochlea (about the size of a pea) that, using different widths and stiffness of the membrane, can differentiate how far along it a traveling wave will go, and so the auditory system responds to different frequencies ending up at different places on the membrane. There is a tonotopic map on the basilar membrane of the frequencies being heard. Further, there is a system of outer hair cells that amplify the inner hair cells. With this the ear can detect frequencies up to 20,000 hertz. A trained musician can distinguish between a tone of 1,000 Hz and another of 1,001 Hz, which requires detecting a difference of only 1 microsecond in the sound wavelengths.

But where is the sound coming from? That too is useful information. Animals have two ears, and the differential travel time of sound from the source to the slightly separated ears can be used to locate the source. But again, this only works in the range 20–2,000 hertz, above which frequency the wavelength is too short to figure location out this way. There is not enough interaural time. So another way is improvised. One ear is in the shadow of the sound, compared to the other. Now the auditory system sends the signals to the superior olive nucleus in the mid-brain, and there the sound from one ear is compared to the sound from the other for the intensity differential resulting from the sound shadow, and the location of higher-frequency sounds is computed. Humans can locate a sound source in the horizontal plane with a precision of 2 degrees (Bear, Conners, and Paradiso 2001:chapter 11). Meanwhile, a spin-off from this auditory system is the vestibular system, used to maintain bodily balance.

One could say that such complex ears were latent in the possibility space of pressure cells, which were latent in the possibility space of carbon, oxygen, nitrogen, and phosphorus atoms. But an equally plausible account is that co-options opened up new possibility space, and the new genetic information achieved proves of value in an evolutionary search for better environmental information. Ears open up the possibility of animal detection of information about their environment, and, in due course, of animal communication.

Anticipating developments during the third big bang, with continuing co-option, much later, ears developed to make possible human language, which makes culture possible, with its cumulative transmission of ideas orally communicated from mind to mind. Escalating co-option drives the information explosion. There are critical turning points in the history of life that hinge on events more idiographic (unique, one-off events) than nomothetic (lawlike, inevitable, repeatable trends). The main idea in co-option is the unpredictable and unexpected; co-option is as revolutionary as it is evolutionary.

Evolutionary Headings: Surprising or Inevitable?

Is evolution going anywhere? If so, are these origins and headings surprising or inevitable? Reflecting on the second big bang, we must return to the same sort of questions about which we puzzled with the first big bang, but now there are more complex dimensions. Something seems to be introducing, layer by layer, new possibilities of order, not just unfolding some latent order already there in the startup. The biological constructions are historical, but they are not simply linear combinatorial processes. True, in the DNA molecules the coding is linear, and the changes are incremental in the linear sequences. But these changes also involve reassorting blocks that reshuffle to produce surprises. A few changes in the linear sequence, resulting from mutations, produce quite different folding patterns at tertiary and quaternary levels in the finished protein (Perutz 1983). Novel possibilities open up whole new regions of search space; old molecules recombine to learn new tricks in unprecedented circumstances. Evolution improvises, sometimes with serendipity.

Evolutionists can make *ex post facto* explanations. After the events have taken place, the paleontologists can say, well, this is what happened, and this is what resulted. But prior to the events, if asked what would be the result if such and such happened, one could not always, from the knowledge of the constituent parts, predict in advance what the results will be. Much less could one predict that such results had to happen. Perhaps one will say, since it has so often happened in evolutionary history, that there must be some disposition in biological nature to improvise, to be opportunistic, some tendency to co-opt.

But where is such tendency located? Hardly from "bottom up" in the precursor materials, inherited from the first big bang, even with their atomic and chemical possibilities. Hardly either from "top down" in the planetary geological or meteorological systems, likewise inherited from inanimate nature. Ecologists think that environments stimulate speciation. The most obvious cybernetic process lies in genetics, so maybe the possibilities lie in the mid-scale genetics. Steven M. Stanley, in his survey of evolutionary speciation, concludes that even more than being environmentally driven, "The system has been essentially internally driven." He finds, with emphasis, "*evidence of great resilience of the intrinsic rates of origination and extinction that characterize individual taxa*" (Stanley 2007:3, 31). The motor of change is not simply challenging environments but prolife organismic drive. Survival of the fittest drives arrival of the more fit, which drives escalating biodiversity and biocomplexity.

So is Darwinian natural selection a sufficient explanation of the emergence of complexity over time? Darwinian natural selection once did not exist; it also had to emerge, at the origin of life. Once under way, across evolutionary natural history the innovations have never violated natural selection; there is always (in principle at least) an explanation in terms of natural selection—at least retrospectively. But prospectively, might the origin and evolution of life have been predicted? Elements of chance and drift appear, and natural selection itself involves elements of randomness, as in mutations. Contemporary biologists are divided across a spectrum whether this creative cybernetic evolutionary history is entirely contingent or quite probable, even inevitable. Francis Crick, after reflecting that the origin of life seems "almost a miracle," continues: "At the present time we can only say that we cannot decide whether the origin of life was an extremely unlikely event or almost a certainty—or any possibility in between these two extremes" (Crick 1981:88).

At one end, famously, Jacques Monod, Nobel prize-winning biologist, insists: "Chance *alone* is at the source of every innovation, of all creation in the biosphere." Evolutionary history is "the product of an enormous lottery presided over by natural selection, blindly picking the rare winners from among numbers drawn at utter random" (Monod 1972:112, 138). That is natural selection with unpredictable results. Similarly and equally famously, Stephen Jay Gould, Harvard paleontologist, looking at the Cambrian explosion, particularly as recorded in fossils found in the Canadian Burgess Shale, concludes: "Almost every interesting event of life's history falls into the realm of contingency" (Gould 1989:290). "We are the accidental result of an unplanned process . . . the fragile result of an enormous concatenation of improbabilities, not the predictable product of any definite process" (Gould 1983:101–102). Life evolves by stumbling around.

Others argue that the chance that complex systems evolve only by chance is vanishingly small. There may be randomness; there is also selection. But of what kind? Simon Conway Morris, eminent Cambridge University paleontologist who did the detailed work on the fossil animals in the Burgess Shale that Gould uses, draws conclusions that are the "exact reverse" (Conway Morris 2003:283). This is one of the more philosophically remarkable happenings in contemporary paleontology. We almost get slapped in the face with what radically different metaphysical frameworks eminent biologists (Harvard versus Cambridge, in this case) can read into, or out of, the same evolutionary facts (Conway Morris and Gould 1998).

Contingency disappears, Conway Morris argues, when we look at the remarkable convergences that have characterized evolutionary history. For example, marsupials evolved in Australia parallel to the evolution of placentals worldwide, some doglike, some catlike, some rodentlike. (Though here we might wonder why kangaroos and wallabies, quite successful marsupials, have no analogs among the placentals.) Eyes, ears, legs, wings appeared more than once. Some species of birds got smarter, so did some species of primates, with very different brain anatomies. If evolution on Earth happened all over again, life would begin in the sea and move to land. There would be plants and animals, predators and prey, genetic coding, sexuality. Sentience would appear in some forms, based on something like neurons, and some of these sentient forms would become increasingly intelligent. So, whatever the contingencies, there must be selection for such events.

Looking back across Earth's natural history and wondering if things might have been otherwise, searching the possibilities for "evolutionary counterfactuals," Conway Morris concludes: "possibly . . . we shall discover in the end that there are none. And, despite the almost crass simplicity of life's building blocks, perhaps we can discern inherent within this framework the inevitable and pre-ordained trajectories of evolution?" (Conway Morris 2003:24). "The details of convergence actually reveal many of the twists and turns of evolutionary change as different starting points are transformed towards common solutions via a variety of well-trodden paths" (Conway Morris 2003:144). Interestingly now, despite the impressively escalating biodiversity, there is something of a counterclaim. The search space, the maneuvering space, is limited; there are only so many niches to occupy. There are constraints, a limited number of best solutions to problems that organisms repeatedly face, so species will often converge. But the constraints still seem to lead upward.

Christian de Duve, also a Nobel prize-winning biologist, finds "landmarks on the pathways of life" that he calls "singularities," the major changes in the history of life (de Duve 2005). He is not thinking only of one-off anomalous events, as others may who use that word. He rather finds the repeated returning of constraints that force a channeling that escalates evolutionary constructions: bottlenecks, environmental constraints, morphological constraints, enzyme functions, molecular structures, "similar singularities," we might call them, as well as, sometimes, what he calls "fantastic luck." These unique features stack up again and again to push up and up, converting chance to the near certainty of startling creativity on Earth, not only in origins but also in trajectories.

de Duve concludes:

> Life was bound to arise under the prevailing conditions, and it will arise similarly wherever and whenever the same conditions obtain. There is hardly any room for "lucky accidents" in the gradual, multistep process whereby life originated. . . . I view this universe [as] . . . made in such a way as to generate life and mind, bound to give birth to thinking beings. (de Duve 1995:xv, xviii)

> It is self-evident that the universe was pregnant with life, and the biosphere with man. (de Duve 2002:298)

de Duve replies to Jacques Monod's claims about life originating only by chance (cited earlier): "To Monod's famous sentence: 'The universe was not pregnant with life, nor the biosphere with man,' I reply: 'You are wrong. They were'" (de Duve 1995:300). "The universe has given life and mind. Consequently, it must have had them, potentially, ever since the Big Bang" (de Duve 2002:298). Again we find prominent biologists in radical disagreement.

Evolutionists here often appeal to deep time, over which eons the unlikely may become likely, even inevitable. Longer time spans do not necessarily help to make the improbable probable, especially where the breakdown rate of the novel constructions always overwhelms the construction rate. This is often problematic when the promising evolutionary constructions are unstable or metastable (temporarily stable). Nevertheless, deep time does give more opportunity for convergence, constraints, channeling to become likely. Rather like cosmologists who, encountering the first big bang, posit a host of multiple universes so that this random one can be less surprising, now the evolutionary biologists posit a host of random mutations endlessly testing life possibilities, and find the emergence and escalation of life on Earth less surprising.

But do such convergences and constraining singularities add up to making the whole life story more or less inevitable? Within the cell Conway Morris notices "some of the proteins being recruited in quite surprising ways from some other function elsewhere in the cell" (Conway Morris 2003:111). "Evolution is a past master at co-option and jury-rigging: redeploying existing structures and cobbling them together in sometimes quite surprising ways. Indeed, in many ways that *is* evolution" (Conway Morris 2003:238). But now it seems not so much that constraints force convergence as, rather, that co-option finds surprising ways to get past earlier constraints. Radical new possibilities open up. Here Conway Morris backs off: "Hindsight and foresight are strictly forbidden . . . we can only retrospect and not predict" (Conway Morris 2003:11–12). So can he, after all, "discern inherent . . . the inevitable and pre-ordained trajectories of evolution?" (Conway Morris 2003:24).

Some events are "quite surprising" indeed. About 2.7 billion years ago (Ga) eucaryotes developed from the ongoing procaryote line. This required radical reorganization of the cell, and seems to have happened only once. Nucleated cells are typically ten times larger in diameter and a thousand, even ten thousand, times larger in volume than nonnucleated cellular organisms. By some accounts this was as innovative as any other discovery in the history of life, because, by concentrating the genetic material, this launched radical new cybernetic possibilities for its elaboration. Christian de Duve traces "this astonishing evolutionary journey." "The consequences of this event were truly epoch-making. . . . Without the emergence of eucaryotic cells, the whole variegated pageantry of plant and animal life would not exist, and no human would be around to enjoy that diversity and to penetrate its secrets" (de Duve 1996:50).

Much later, but before plants and animals had diverged, by endosymbiosis what were once-independent organisms fused into other, larger and quite different organisms to become mitochondria transferred into the pre-plant/animal line, and became the powerhouse organelles for all subsequent life (fig. 2.12; data from Dyall, Brown, and Johnson 2004).

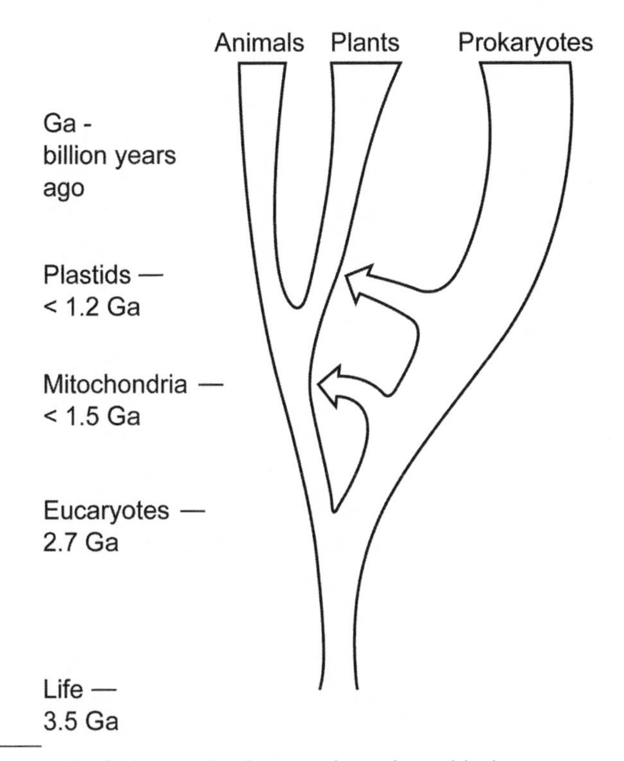

Figure 2.12 Evolutionary development by endosymbiosis

There emerged a new kind of system where the organism has highly efficient and specialized power modules (the mitochondria), something not possible for either of the precedents before they interacted, criss-crossed, synthesized, and transformed each other. The "information" about how to do this was not present before in the preceding organisms, but now there appears new "information" (coded in the revised DNA in the nucleus and the residual DNA in the mitochondrion) that makes this new, high-powered form of life possible.

About 1.6 billion years ago the plant and animal lines diverged; and later still, by another remarkable endosymbiosis this time plastids, once free-living, made the lateral transfer into the plant line to become the chloroplasts critical for the capture of solar energy. Again, new, higher-powered forms of life became possible, both in the plants and in the animals that feed on plants (Dyall, Brown, and Johnson 2004). Perhaps one can say that endosymbiosis is likely to occur: there are frequently "mobile elements" that transpose and reshape evolution (Kazazian 2004). But is there any "inherency" in the earliest microbial life making inevitable or even probable these two especially vital endosymbioses, both thought to have initiated as singularities and both dramatically changing the history of life on Earth?

One can say that evolution is disposed to exciting serendipity. In such cases of co-opted emergence, repeatedly compounding, something that is genuinely new pops out, pops up. The novelty is, of course, based on the precedents, but there is genuine novelty not present in any of the precedents. The presence of the prior organisms was required, but did not determine or make inevitable these results.

Conway Morris and de Duve are swimming upstream against a powerful current in contemporary theoretical biology. As noted earlier, considering escalating information transfer, John Maynard Smith and Eörs Szathmáry found major escalations at critical thresholds in evolution: the origin of the genetic code, the origin of eucaryotes from procaryotes, meiotic sex, multicellular life, animal societies, and human language. But they find "no reason to regard the unique transitions as the inevitable result of some general law"; to the contrary, these events might not have happened at all (Maynard Smith and Szathmáry 1995:3). Both Maynard Smith and de Duve find a series of escalation events, although Maynard Smith finds them unexpected; de Duve finds them expected.

Was all that resulted already there in the potential at the big bang, inevitably to unfold and evolve, as de Duve claims? Were later-coming fauna and flora, including humans, already there in the possibility spaces of even the simplest predecessor organisms? Maybe some of the possibility was within one organism, some within another. This seems to argue that whatever eventuates must always have been potentially there in the primordial matter-energy. Conway Morris suspects that the basic forms into which living organisms can evolve are preset; "the nodes of occupation are effectively predetermined from the Big Bang (Conway Morris 2003:310). That is a metaphysical claim, not a scientific claim.

But isn't it equally plausible to believe that new possibility space opens up, not previously there? New information, as in DNA, opens up new opportunities, previously impossible to hydrogen, carbon, oxygen, iron so long as they are devoid of such information. New possibility space appeared with the co-option of the mitochondria and chloroplast predecessor organisms to novel functions. Some achievements that are genuinely new pop up. The precedents in both their actuality and possibilities are necessary but not sufficient for the consequents. There is breakthrough discovery, innovative creativity.

Biologists, a century back, used to call such events "saltations." Physicists, pressed for words from their discipline, might call it a "quantum leap." Maybe we need a new term: "cybernetic leap." Biologists inclined toward chance may call this "tinkering" (Jacob 1977). Biologists impressed with the novel results will call it evolutionary "exploring" or "searching." Some call it "genetic engineering" (Shapiro 1998; 2005). Historians will remark that such events are narrative adventures; they do not follow any Aristotelian logic, nor any hypothetico-deductive science. One needs a metaphysics for such co-option because there appear new ontological levels, both actual and possible. Sight appears where before was only heat-stress protection, language where before was only skin-pressure sensibility. Sight and language open up the possibility of writing/reading. Co-option is a vital key to historical creativity.

Retrospectively, of course, after these novelties happen, the historian can trace the steps by which they happened. The paleontologist and paleo-molecular biologists can give scientific explanations, a posteriori. But at each developmental juncture, were (*per impossible*) a biologist standing there watching, nothing would be a priori. One can claim that the possibilities were always there; one can with equal plausibility claim that new possibility space has opened up en route in the course of natural history. Prospectively, if one could stand at each present moment, at each "now" over the course of evolution, there would always be the great unknown. There is the generation of new possibility space in which information breakthroughs become possible. The pivotal element in a metaphysics of such evolutionary biology is the future—not the past, not even the present. Past and present are necessary but never sufficient for the future. In that sense our accounts will always be insufficient, incomplete, before this capacity for future innovation.

Sometimes the explanatory account is by laws applied to initial conditions, and the same laws again reapplied to the resulting outcomes, now treated as further initial conditions. But sometimes, with co-options, endosymbioses, lateral genetic transfers, mutations, the outcomes are not just further sets of initial conditions. The novel outcomes revise the previous laws; the rules of the game change, as well as the initial conditions, and the future is like no previous past. One can say that all this surprising serendipity is somehow "inherent" from the start; but the explanatory power of such a claim is rather vague. Predictably, there will be unpredictable co-options!

Stuart Kauffman puts this pointedly:

> Now the critical question. Do you think you could say ahead of time, or finitely prestate, all possible Darwinian preadaptations of, say, species alive now, or even humans? . . . Ever novel functionalities come to exist and proliferate in the biosphere. The fact that we cannot prestate them is essential, and an essential limitation to the way Newton taught us to do science. Prestate the relevant variables, the forces acting among them, and the initial and boundary conditions, and calculate the future of the system. . . . We are profoundly precluded from the Newtonian move. In short, the evolution of the biosphere

> is radically unknowable. . . . The evolution of the biosphere is radically creative, ceaselessly creative, in ways that cannot be foretold. (Kauffman 2007:912–913)

Astronomers speculate about their multiverses; biologists celebrate on this Earth alone rich plurality of kinds, an endlessly creative explosion with a radically open future—unless, alas, the species appearing at the third big bang puts this life on Earth in jeopardy. Reflecting about the first big bang, we observed that a theory of everything that made the anthropic qualities of matter-energy inevitable would still require a deeper account of why there should be such a universe. Similarly, a surprising universe of the kind we have needs a deeper account. We reach the same puzzle here on Earth below that we found in the heavens above. If life is inevitable, it is remarkable. If life is contingent, it is equally remarkable. Either way, there is radical creativity demanding a deeper account.

The predictable/unpredictable debate couples with further debate about whether to call what has happened progressive change. Again, biologists and philosophers of biology are quite divided. Michael Ruse insists, "Evolution is going nowhere—and rather slowly at that" (Ruse 1986:203). A frequent argument is that most forms of life, although they may respeciate and differ, do not get any smarter—the beetles or the plants. The linchpin of contemporary biology is that the better adapted survive, but the better adaptations with which most species survive have nothing to do with evolutionary progress—those beetles and plants again. Anyone who today believes that progress was a heading during evolutionary history, Ruse concludes, is guilty of "pseudo-science." Trying to document this in *Monad to Man*, a 400-page Harvard University Press book, Ruse himself goes rather slowly, and one reason is that he has to argue away what many classical biologists have believed: that there is some tendency toward increased complexity across the millennia of natural history, and that this is some sort of advance.

Ernst Mayr asks:

> Who can deny that overall there is an advance from the procaryotes that dominated the living world more than three billion years ago to

> the eucaryotes with their well organized nucleus and chromosomes as well as cytoplasmic organelles; from the single-celled eucaryotes to metaphytes and metazoans with a strict division of labor among their highly specialized organ systems; within the metazoans from ectotherms that are at the mercy of climate to the warm-blooded endotherms, and within the endotherms from types with a small brain and low social organization to those with a very large central nervous system, highly developed parental care, and the capacity to transmit information from generation to generation? (Mayr 1988:251–252, 256)

Indeed, the series of morphological and physiological innovations that have occurred in the course of evolution can hardly be described as anything but progress. (Mayr 1982:532)

Edward O. Wilson concludes his study of the diversity of life:

> Biological diversity embraces a vast number of conditions that range from the simple to the complex, with the simple appearing first in evolution and the more complex later. Many reversals have occurred along the way, but the overall average across the history of life has moved from the simple and few to the more complex and numerous. During the past billion years, animals as a whole evolved upward in body size, feeding and defensive techniques, brain and behavioral complexity, social organization, and precision of environmental control—in each case farther from the nonliving state than their simpler antecedents did.
>
> More precisely, the overall averages of these traits and their upper extremes went up. Progress, then is a property of the evolution of life as a whole by almost any conceivable intuitive standard, including the acquisition of goals and intentions in the behavior of animals. It makes little sense to judge it irrelevant. . . . In spite of major and minor temporary setbacks, in spite of the nearly complete turnover of species, genera, and families on repeated occasions, the trend toward biodiversity has been consistently upward. (Wilson 1992:187, 194)

George Gaylord Simpson, after surveying the fossil record extensively and noting that there are exceptions, concludes, "The evidence warrants considering general in the course of evolution . . . a tendency for life to expand, to fill in all available spaces in the liveable environments, including those created by the process of that expansion itself. . . . The total number and variety of organisms existing in the world has shown a tendency to increase markedly during the history of life" (1967:242, 342). R. H. Whittaker finds, despite "island" and other local saturations and equilibria, that on continental scales and for most groups "increase of species diversity . . . is a self-augmenting evolutionary process without any evident limit." There is a natural tendency toward increased "species packing" (1972:214).

William Day concludes that "as we arrange the sequences of evolution's advance, we discover an unsettling implication":

> Each step is an evolutionary curve; all steps together outline an accelerating advance for all biological evolution. . . . Each major development in evolution appears to take less and time to occur. And each development begins slowly but, fed by its own momentum, begins to accelerate until it races to its developed state. When it reaches a final level—a higher stage in evolution—the offspring of the new life form begin to repeat the cycle, evolving some feature that ultimately leads to another succeeding step . . . it continues to accelerate stage after stage. . . . We are in the middle of something momentous that is taking place. (Day 1984:257–258)

Karl Popper concludes that science discovers "a world of propensities," open to historical innovation, the possibility space ever enlarging.

> In our real changing world, the situation and, with it, the possibilities, and thus the propensities, change all the time. . . . This view of propensities allows us to see in a new light the processes that constitute our world: the world process. The world is no longer a causal machine—it can now be seen as a world of propensities, as an unfolding process of realizing possibilities and of unfolding new possibilities. . . . New possibilities are created, possibilities that previously simply did not exist. . . . Especially in the evolution of biochemistry, it is widely appreciated that every new compound

> creates new possibilities for further new compounds to synthesize: possibilities which previously did not exist. The possibility space . . . is growing. . . . Our world of propensities is inherently creative. (Popper 1990:17–20)

The result is the evolutionary drama. "The variety of those [organisms] that have realized themselves is staggering." "In the end, we ourselves become possible" (1990:26, 19).

Stuart Kauffman similarly finds himself amazed at life, at agency: "It is utterly remarkable that agency has arisen in the universe—systems that are able to act on their own behalf; systems that modify the universe on their own behalf. Out of agency comes value and meaning." "Life is valuable on its own, a wonder of emergence, evolution and creativity. Reality is truly stunning" (Kauffman 2007:909, 914).

Like the theory of everything, which would make cosmic origins inevitable, we do not know if we will get a parallel theory of inevitable evolution. If the scientists say "contingent" evolution, metaphysicians will likely say that this is not enough explanation for the second big bang and its three-billion-year explosive history, any more than "random" is enough explanation for the first big bang and its results. If the scientists say "inevitable" evolution, the metaphysicians will still say that such events demand further explanation, just as does a theory of everything that would make the first big bang inevitable. If, as seems more likely, biologists find some mixing of the two, the results on Earth are still impressive, indeed explosive. Both historically and logically, the generative context of life requires both order and contingency. We found earlier that exploring immense possibilities in the life struggle cannot be fine-tuned clockwork.

Evolutionary tendencies go beyond optimizing local species survival, exploring combinatorial state spaces that increase exponentially with the number of components. Among these, brains are especially impressive. Life starts up, and, on many of its trajectories, it smarts up. This exuberant proliferation of life on Earth is a second big bang, demanding a philosophical response in the only one of Earth's creatures capable of such a response. But to get such cerebral capacities we need a third big bang.

CHAPTER 3

Mind

The Human Big Bang

The third big bang is the explosive growth of the human brain, sponsoring the human mind. Uniquely among the species on Earth, *Homo sapiens* is cognitively spectacular. Consider a graph of increasing cranial capacity in the hominoid line (fig. 3.1, Pilbeam 1972/ Wilson 1975). That trajectory makes virtually a right-angle turn. Edward O. Wilson remarks: "No organ in the history of life has grown faster" (Wilson 1978:87). Steve Dorus and his team of neurogeneticists conclude: "Human evolution is characterized by a dramatic increase in brain size and complexity" (Dorus et al. 2004:1027). Theodosius Dobzhansky puts it this way: "The biological evolution has transcended itself in the human 'revolution'" (Dobzhansky 1967:58).

J. Craig Venter and more than 200 coauthors call the human brain "a massive singularity." Reporting on the completion of the Celera Genomics version of the human genome project, they caution in their concluding paragraph:

> In organisms with complex nervous systems, neither gene number, neuron number, nor number of cell types correlates in any meaningful manner with even simplistic measures of structural or behavioral

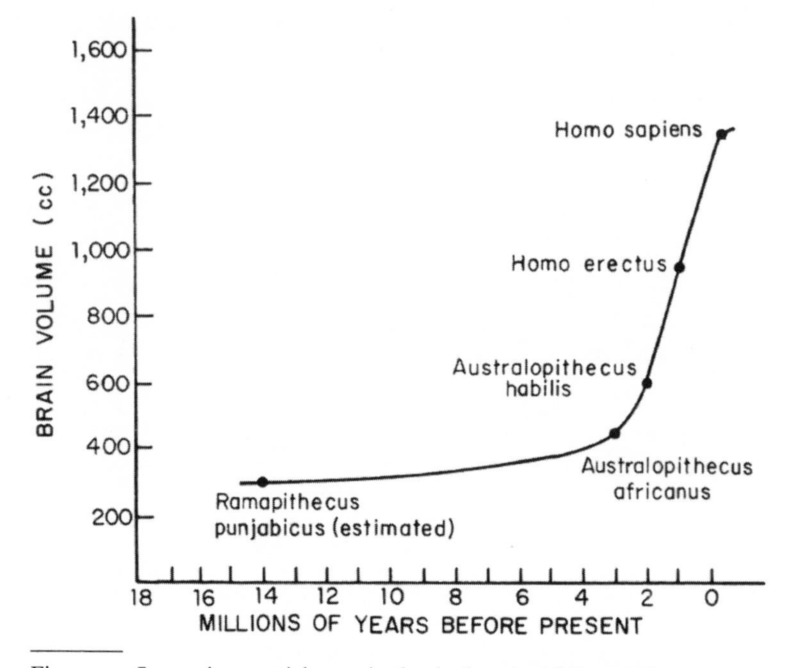

Figure 3.1 Increasing cranial capacity in the hominoid line (*Wilson* 1975, *from Pilbeam* 1972)

> complexity. . . . Between humans and chimpanzees, the gene number, gene structures and functions, chromosomal and genomic organization, and cell types and neuroanatomies are almost indistinguishable, yet the development modifications that predisposed human lineages to cortical expansion and development of the larynx, giving rise to language, culminated in a massive singularity that by even the simplest of criteria made humans more complex in a behavioral sense. . . . The real challenge of human biology, beyond the task of finding out how genes orchestrate the construction and maintenance of the miraculous mechanism of our bodies, will lie ahead as we seek to explain how our minds have come to organize thoughts sufficiently well to investigate our own existence. (Venter et al. 2001, 1347–1348)

The human embryo starts as a single cell, without consciousness, and in the adult human an ideational consciousness is life-orienting. This "massive singularity" these geneticists find in humans is all the more interesting when it comes by consensus of the same geneticists discovering the genetic continuities with the other primates from which humans originated.

Bruce Lahn, the lead researcher in the Dorus study, concluded:

> We've proven that there is a big distinction. Human evolution is, in fact, a privileged process because it involves a large number of mutations in a large number of genes. . . . To accomplish so much in so little evolutionary time—a few million years—requires a selection process that is perhaps categorically different from the typical processes of acquiring new biological traits. . . . It required a level of selection that is unprecedented. Our study offers the first genetic evidence that humans occupy a unique position in the tree of life. Simply put, evolution has been working very hard to produce us humans. (Lahn, interviewed in Gianaro 2005)

David Premack puts this differently but with equal force: "Human language is an embarassment for evolutionary theory because it is vastly more powerful than one can account for in terms of selective fitness" (Premack 1985:282).

Michael Gazzaniga speaks of "the explosion in human brain size":

> We are hugely different. While most of our genes and brain architecture are held in common with animals, there are always differences to be found. And while we can use lathes to mill fine jewelry, and chimps can use stones to crack open nuts, the differences are light years apart. . . . A phase shift occurred. (Gazzaniga 2008:13, 3)

So the puzzle is how can a change of some one or two percent in DNA results in light-years of mental explosion.

Theory of Mind: The Human Singularity

Perhaps the radical threshold is crossed with the emergence of consciousness, already present for millennia in the primate line. Certainly, consciousness prior to humans is an already radical emergent. Such escalating psyche has been impressive, resulting from the second big bang. Still, such primary consciousness was graded and simplified through earlier species, present in much simpler-brained organisms: birds, reptiles, fish, invertebrates; we do not know how or where to draw the lines. Consciousness, however novel, did gradually evolve. "Conscious" has the root meaning "I know."

Although consciousness long preceded humans, there was an explosive state change when humans crossed a divide and gained their self-reflexive, ideational, linguistic, symbolic capacities. Humans are not simply conscious; they are self-conscious. The human brain is not just a scaled-up version of chimpanzee brain. Humans are remarkable in their capacities to process thoughts, ideas, symbolic abstractions figured into interpretive gestalts with which the world is understood and life is oriented. This higher consciousness is a constitutive dimension of humans absent in all other species. "I know that I know."

We do not know with precision when this took place; probably over millennia. The key threshold is the capacity to pass ideas from mind to mind. There is no clear evidence that chimpanzees attribute mental states to others. Chimps have little or no "theory of mind"; they do not know other minds are there with whom they might communicate, to learn what they know. Or, if you prefer to say that one chimp can know what another knows, chimps have a theory of immediate mind (one chimp sees that another chimp knows where those bananas are); humans have a theory of the ideational mind (one human teaches another the Pythagorean theorem). Humans have ideational uniqueness, further examined below.

Animals are variously socialized, and become what they become interactively with their surroundings, which include the groups in which they live. But there is little or no evidence for any higher-order intentionality, even among primates that are highly social. Organisms with zero-order intentionality have no mental beliefs or desires at all. (We did find in the previous chapter that some geneticists can speak of genetic intention.) Animals clearly intend to change the behavior of other animals, first-order intentionality. Second-order intentionality would involve intent to change the mind, as distinguished from the behavior (though perhaps the behavior as well) of another animal. Third-order intentionality would involve one's knowledge that another mind is intending to change one's mind (Dennett 1987). Primates do not seem to realize that there are minds present to teach in others, although they often imitate each other's behavior, as when adults are imitated by their offspring.

In this higher-order sense of communication, conclude Dorothy L. Cheney and Robert M. Seyfarth, "signaler and recipient take into account each others' states of mind. By this criterion, it is highly doubtful that *any* animal signals could ever be described as truly communicative." They continue:

> It is far from clear whether any nonhuman primates ever communicate with the intent to inform in the sense that they recognize that they have information that others do not possess. . . . There is as yet little evidence of any higher-order intentionality among nonhuman species. (Cheney and Seyfarth 1990:142, 209)

Although chimpanzees collaborate to hunt or get food, Michael Tomasello and his colleagues conclude, "it may be said with confidence that chimpanzees do not engage in collaborative learning. . . . They do not conceive of others as reflective agents—they do not mentally simulate the perspective of another person or chimpanzee simulating their perspective. . . . There is no known evidence that chimpanzees, whatever their background and training, are capable of thinking of other interactants reflectively" (Tomasello et al. 1993:504–505).

Daniel Povinelli and his colleagues conclude of chimps: "There is considerable reason to suppose that they do not harbor representations of mental states in general. . . . Although humans, chimpanzees, and most other species may be said to possess mental states, humans alone may have evolved a cognitive specialization for reasoning about such states" (Povinelli, Bering, and Giambrone 2000:509; Povinelli and Vonk 2003). "Humans have a whole system that we call theory of mind that chimps don't have" (Povinelli, quoted in Pennisi 1999:2076). Carl Zimmer concludes: "Of all the species on Earth, only humans possess what researchers call a 'theory of mind'—the ability to infer what others are thinking. . . . After decades of studies, no one has found indisputable signs that chimps or other nonhuman primates have a theory of mind." "Understanding what others are thinking is a human exclusive" (Zimmer 2003).

Joaquín M. Fuster, a neuroscientist, finds that in human brains there is an "emergent property" that is "most difficult to define":

> As networks fan outward and upward in associative neocortex, they become capable of generating novel representations that are not reducible to their inputs or to their individual neuronal components. Those representations are the product of complex, nonlinear, and near-chaotic interactions between innumerable elements of high-level networks far removed from sensory receptors or motor effectors. Then, top-down network building predominates. Imagination, creativity, and intuition are some of the cognitive attributes of those emergent high-level representations. (Fuster 2003:53)

When one is imagining, daydreaming (or dreaming), the mental activity going on is remote from actual perception or bodily behaviors. Even if one is thinking about future behavior, there can be complex ideational structure remote from immediate experience. "Molybdenum steel has a higher tensile strength; maybe I should replace that failing crankshaft with better steel. I'll ask Sam what he thinks." Humans live in an ideational world, minds contemplating and contacting other minds. Christopher Frith shares his thoughts: "To get an idea from one brain into another, that's a deeply mysterious thing that we do" (quoted in Zimmer 2003).

Hyperimmense Brain: Neural Explosion

Christian de Duve finds the rapid evolution of the human brain "dazzling":

> Most impressive is the development of the brain, which took place at a staggering speed. After having taken some 600 million years

> to reach a volume on the order of 450 cm^3 in our simian ancestors, the size of the hominoid brain went through an astonishingly rapid phase of expansion, virtually jumping to three times this value in a little more than two million years. On the evolutionary time scale, such a rate of change is no less than dazzling. (de Duve 2002:189)

But now the information in DNA, however necessary, proves inadequate. Genes cannot wire up the mature human brain. Impressive though the amount of genetic information in humans is, this is far too little with which to build a functioning human brain. The number of neurons and their possible connections is far vaster than the number of genes coding for the neural system. So it is impossible for the genes to specify all the needed neural connections. The genes in fetus and womb seem to have learned how to generate by repeated algorithms a dynamic and open-ended neural network that, in due course, makes itself. Brain-forming genes do not specify some product with stereotyped function; rather, by splicing and resplicing, cutting and shuffling, the brain genes proliferate cascading neurons with almost endless possibilities of organization, depending on how they synaptically connect themselves up.

Genes create the instruments, but the orchestration is cerebral. The secret of our advanced information lies somewhere else, resulting from genetic flexibility that opens up cerebral capacity. Barry J. Dickson concludes: "The ultimate challenge, after all, is to find out how a comparatively small number of guidance molecules generate such astonishingly complex patterns of neuronal wiring" (Dickson 2002:1963). "Changes in protein and gene expression have been particularly pronounced in the human brain. Striking differences exist in morphology and cognitive abilities between humans and their closest evolutionary relatives, the chimpanzees." So conclude a team of molecular biologists and evolutionary anthropologists from the Max Planck Institutes in Germany (Enard et al. 2002).

Geneticists decoded the human genome, confirming how little humans differ in their protein molecules from chimpanzees, yet simultaneously realizing that the startling successes of humans doing just this sequencing of their own genome as readily proves human distinctiveness. Humans have made an exodus from determination by genetics and natural selection and passed into a mental and social realm with new freedoms. In body structures generally, such as blood or liver, humans and chimpanzees are 95 to 98 percent identical in their genomic DNA sequences and the resulting proteins. But we have over three times their cranial cortex, over 300 percent difference in the head. This cognitive development has come to a striking expression point in the hominid lines leading to *Homo sapiens*, going from about 300 to 450 cubic centimeters of cranial capacity in chimpanzeelike ancestors to 1,400 to 1,500 cc. in humans. Nor is absolute brain size the only consideration; relative brain size is another. There too, relative to our body size, the human brain is proportionally bigger than that of any other animal. Some brains are more convoluted and complex than others of the same size. Neanderthal humans had somewhat larger brains than contemporary humans, though less convoluted.

Animal brains are already impressive. In a cubic millimeter (about a pinhead) of mouse cortex there are 450 meters of dendrites and one to two kilometers of axons. Human brains multiply the cortex in mice 3,000 times. The connecting fibers in a human brain, extended, would wrap around the Earth 40 times.

Geneticists have recently also sequenced the chimpanzee genome; comparing it with the human genome, they are still trying to figure out how so few genetic differences made such an enormous brainpower difference (Chimpanzee Sequencing and Analysis Consortium 2005). Quantitative genetic differences add into qualitative differences in capacity, an emerging cognitive possibility and practical performance that exceeds anything known in previous evolutionary achievements.

Some transgenetic threshold seems to have been crossed. The human brain is of such complexity that descriptive numbers are astronomical and difficult to fathom. A typical estimate is 10^{12} neurons, each with several thousand synapses (possibly tens of thousands). Each neuron can "talk" to many others. The postsynaptic membrane contains more than a thousand different proteins in the signal-receiving surface. "The most molecularly complex structure known [in the human body] is the postsynaptic side of the synapse," according to Seth Grant, a neuroscientist (quoted in Pennisi 2006). More than a hundred of these proteins were co-opted from previous, non-neural uses, but by far most of them evolved during brain evolution. "The postsynaptic complexes and the [signaling] systems have increased in complexity throughout evolution," says Berit Kerner, geneticist at the University of California, Los Angeles (quoted in Pennisi 2006). This is nature's nanotechnology.

This nanophysiology is integrated into a dendritic network structured at multiple hierarchical levels. The nerve dynamics "don't violate the equations of physics and chemistry, but they cannot be derived from them. They are new laws appropriate to the science of electrophysiology, which is removed by several hierarchical levels from atomic physics" and "independent of the equations of physics and chemistry" (Scott 1995:182). This network, formed and reformed, makes possible virtually endless mental activity. Much, even most of what goes on in our brains is below the level of conscious awareness, of course; but humans can bring novel cognitive capacities to critical focus.

The result is a mental combinatorial explosion superimposed not just on the physics and chemistry, but further on the biological combinatorial explosion that we earlier met. The human brain is capable of forming something in the range of $10^{70,000,000,000}$ thoughts—a number that dwarfs the number of atoms in the visible universe (10^{80}) (Flanagan 1992:37; Holderness 2001). On a cosmic scale, humans are minuscule atoms, but on a complexity scale, humans have "hyperimmense" possibilities in mental complexity (Scott 1995:81). In our 150 pounds of protoplasm, in our 3-pound brain is more operational organization than in the whole of the Andromeda galaxy.

Genes make the kind of human brains possible that facilitate an open mind. But when that happens, these processes can also work the other way around. What began as a "bottom-up" process becomes a "top-down" process. In "top-down" causation an emergent phenomenon reshapes and controls its precedents, as contrasted with "bottom-up" causation, in which precedent, simpler causes are fully determinative of more complex outcomes. We encountered this before with organisms interacting with their chemistries. Now we encounter it again, at a higher level.

Minds employ and reshape their brains to facilitate their chosen ideologies and lifestyles. We neuroimage brain blood flow to find that such thoughts can reshape the brains in which they arise. This in turn can affect bodily behavior. Michael Merzenich, a neuroscientist, reports his increasing appreciation of "what is the most remarkable quality of our brain: its capacity to develop and to specialize its own processing machinery, to shape its own abilities, and to enable, through hard brainwork, its own achievements" (Merzenich 2001:418).

In the vocabulary of neuroscience, we have "mutable maps" in our cortical representations, formed and reformed by our deliberated changes in thinking and resulting behaviors. We do require dimensions of our brains that are specified by genetics, as in hearing and seeing. But our brains are also quite plastic, forging properties enabled by our genes but shaped by our experience, environmental and social (neuroplasticity). For example, with the decision to play a violin well, and resolute practice, string musicians alter the synaptic connections and thereby the structural configuration of their brains to facilitate fingering the strings with one arm and drawing the bow with the other (Elbert et al. 1995). Likewise, musicians enhance their hearing sensitivity to tones, enlarging the relevant auditory cortex by 25 percent compared with nonmusicians (Pantev et al. 1998).

With the decision to become a taxi driver in London, and long experience driving in the city, drivers likewise alter their brain structures, devoting more space to navigation-related skills than non-taxi drivers have. "There is a capacity for local plastic change in the structure of the healthy adult human brain in response to environmental demands" (Maguire et al. 2000:4398). Similarly, researchers have found that "the structure of the human brain is altered by the experience of acquiring a second language" (Mechelli et al. 2004). Or by learning to juggle (Draganski et al. 2004).

The human brain is as open as it is wired up. Our minds shape our brains. We form a synaptic self; synapses and experiential self are reciprocal processes. One can say that finding differing locations in the brain where differing kinds of mental activities take place is evidence for the physical basis of our mental activities. This is true. But another way to interpret the same evidence is that our mental decisions to become a violin player or taxi driver, or learn a second language, reallocate brain locations to new functions in support of these decisions.

Violin players, taxi drivers, jugglers use highly localized areas of the brain. But other skills, such as gaining a higher education, are more pervasively distributed. The authors of a leading neuroscience text use the violin players as an icon for us all, and conclude: "It is likely that this is an exaggerated version of a continuous mapping process that goes on in everyone's brain as their life experiences vary" (Bear, Connors, and Paradiso 2001:418). We have no apparatus to measure such more global synaptic changes, but every reason to think there are there (LeDoux 2002).

Neuroscience went molecular (acetylcholine in synaptic junctions, voltage-gated potassium channels triggering synapsizing) to discover that what is really of interest is how these synaptic connections are configured by the information stored there, enabling function in the inhabited world. Our ideas and our practices configure and reconfigure our own sponsoring brain structures.

Ideational Uniqueness: Cultural Explosion

What is missing in the primates is precisely what makes a cumulative, transmissible human culture possible. The central idea is that acquired knowledge and behavior is learned and transmitted from person to person, by one generation teaching another. Ideas pass from mind to mind, in large part through the medium of language, with such knowledge and behavior resulting in a greatly rebuilt, or cultured, environment. Andrew Whiten finds:

> When we focus our comparative lens on culture, the evidence is all around us that a gulf separates humans from all other animals. . . . Ape culture may be particularly complex among non-human animals, yet it clearly falls short of human culture. An influential contemporary view is that the key difference lies in the human capacity for cumulative culture. . . . [In chimps,] hints of cumulation exist, such as the refinement of using prop stones to stabilize stone anvils during nut cracking, but these remain primitive and fleeting by human standards. (Whiten 2005:52–53)

Humans live under what Robert Boyd and Peter J. Richerson call "a dual inheritance system," both genes and culture (Boyd and Richerson 1985). They find "that the existence of human culture is a deep evolutionary mystery on a par with the origins of life itself." "Human societies are a spectacular anomaly in the animal world" (Richerson and Boyd 2005:126, 195). The human transition into culture is exponential, nonlinear, reaching extraordinary epistemic powers. To borrow a term from the geologists, humans have crossed an unconformity. To borrow from classical philosophers, we are looking for the unique *differentia* of our *genus*.

What is quite surprising in humans is not so much that they have intelligence generically, for many other animals have specific forms of a generic intelligence. Nor is it that humans have intelligence with subjectivity, for there are precursors of this too in the primates. The surprise is that this intelligence becomes reflectively self-conscious and builds cumulative transmissible cultures. An information explosion gets pinpointed in humans.

The variation on which selection acts does not arise in the genes but in the mind, ideational variation, not mutations in DNA. The selection, if it remains at times natural selection (more offspring in the next generation) passes over into ideational, cultural selection (Einstein over Newton, Jesus transforming Judaism). The evolved brain allows many sets of mind: one does not have to have Plato's genes to be a Platonist, Darwin's genes to be a Darwinian, or Jesus' genes to be a Christian. The system of inheritance of ideas is independent of the system of inheritance of genes. Ideas can jump across genetic lines.

The determinants of animal and plant behavior are never anthropological, political, economic, technological, scientific, philosophical, ethical, or religious. The intellectual and social heritage of past generations, lived out in the present, reformed and transmitted to the next generation, is regularly decisive in culture. "Culture," by Margaret Mead's account, is "the systematic body of learned behavior which is transmitted from parents to children" (1989:xi). Culture, according to Edward B. Tylor's classic definition, is "that complex whole which includes knowledge, belief, art, morals, law, custom, and any other capabilities and habits acquired by man as a member of society" (1903:1).

Animal ethologists have complained that such accounts of culture are too anthropocentric and need to be more inclusive of animals (de Waal 1999). Partly because of new animal behaviors observed, but mostly by enlarging (or, if you like, shrinking) the definition, it has become fashionable to claim that animals have culture. Although finding human culture a "spectacular anomaly," still Robert Boyd and Peter J. Richerson are willing to revise the definition of culture: "Culture is information capable of affecting individuals' phenotypes which they acquire from other conspecifics by teaching or imitation" (1985:33). The addition of "imitation" greatly expands and simultaneously dilutes what counts as culture. Animals may observe and learn. By this account there is culture when apes "ape" each other, but also culture in horses and dogs, beavers, rats—wherever animals imitate the behaviors of parents and conspecifics. Geese, with a genetic tendency to migrate, learn the route by following others; warblers, with a tendency to sing, learn to sing better when they hear others. Copied song dialects may persist over several generations. Whales and dolphins communicate by copying the noises they hear from others; this vocal imitation constitutes culture at sea (Rendell and Whitehead 2001).

But with "culture" extending from people to warblers, "culture" has become a nondiscriminating category for the concerns we here wish to analyze. One finds widespread animal cultures by lowering the standards of evidence. Critical to a more discriminating analysis is the difference between mind-mind interactions, the sharing of ideas, pervasive in human cultures, and not mere behavioral imitation, copying what another does, which is widespread among animals that can acquire information. If we are going to call what warblers and geese do "culture," then we will need to invent another word, "superculture," to describe what humans do, which is indeed "super" to these animal capacities.

Opening an anthology on *Chimpanzee Culture*, the authors doubt, interestingly, whether there is much of such a thing: "Cultural transmission among chimpanzees is, at best, inefficient, and possibly absent." There is scant and in some cases negative evidence for active teaching of the likeliest features to be transmitted, such as tool-using techniques. Chimpanzees clearly influence each other's behavior, and seem to intend to do that; they copy the behavior of others. But there is no clear evidence that they attribute mental states to others. They seem, conclude these authors, "restricted to private conceptual worlds" (Wrangham et al. 1994:2). Christophe Boesch finds population-specific behaviors in chimpanzees, which spread by imitation, and is willing to call this "culture," but he adds: "It seems far-fetched to pretend that human cultures are similar to chimpanzee culture" (Boesch 1996:266).

Without some concept of teaching, of ideas moving from mind to mind, from parent to child, from teacher to pupil, a cumulative transmissible culture is impossible. Humans learn what they realize others know; they employ these ideas and resulting behaviors; they evaluate, test, and modify them, and, in turn, teach what they know to others, including the next generation. So human cultures cumulate, but with animals there is no such cultural "ratchet" effect.

In a lead article in *Behavioral and Brain Sciences,* Michael Tomasello, Ann Cale Kruger, and Hilary Horn Ratner pinpoint this difference:

> Simply put, human beings learn from one another in ways that nonhuman animals do not. . . . Human beings are able to learn from one another in this way because they have very powerful, perhaps uniquely powerful, forms of social cognition. Human beings understand and take the perspective of others in a manner and to a degree that allows them to participate more intimately than nonhuman animals in the knowledge and skills of conspecifics. (Tomasello, Kruger, and Ratner 1993:495)

Tomasello continues: "Nonhuman primates in their natural habitats . . . do not intentionally teach other individuals new behaviors" (Tomasello 1999:21).

Cheney and Seyfarth report from their studies of monkeys: "Teaching would seem to demand some ability to attribute states of mind to others. . . . Even in the most well documented cases, however, active instruction by adults seem to be absent" (Cheney and Seyfarth 1990:223–225). Tetsuro Matsuzawa concludes: "There is no overt teaching behavior in chimpanzees. . . . Chimpanzee mothers in the wild do not teach as human mothers do" (Matsuzawa 2007:99–100).

Bennett G. Galef Jr. concludes: "As far as is known, no nonhuman animal teaches" (Galef 1992:161). "Given that imitation is rare in nonhuman primates and teaching is essentially nonexistent, it's hard to see how you are going to get the cumulative culture which is the hallmark of our culture" (Galef, quoted in Vogel 1999:2072).

Animals do not have a sense of mutual gaze as joint attention, of looking *with*. "Nonhuman primates in their natural habitats . . . do not point or gesture to outside objects for others; do not hold objects up to show them to others; do not try to bring others to locations so that they can observe things there; do not actively offer objects to other individuals by holding them out" (Tomasello 1999:21). Animals do see others in pursuit of the food, mates, or territories they wish to have; but they do not know that other minds are there to teach.

One can trim down the meaning of "teaching," somewhat similarly to reducing the definition of "culture," and find noncognitive accounts of teaching. Interestingly, a recent study suggests a form of teaching not in the primates, where it is usually looked for, but in wild meerkats. Adults differentially cripple prey for their young to hunt, depending on how naïve the juvenile hunter is. They cripple scorpions, prey with dangerous stingers, differently depending on how adept the juvenile is at handling them (Thornton and McAuliffe 2006). Many predators release crippled prey before their young, encouraging their developing hunting skills (Caro and Hauser 1992).

But if teaching is found anywhere individuals have learned to modify their behavior with the result that the naïve learn more quickly, then teaching is found in chickens in the barnyard, when a mother hen scratches and clucks to call her chicks to newfound food and the chicks soon imitate her. The meerkat researchers conclude that there is only simple differential behavior responding to the handling skills of the pups, without the presence of ideas passing from mind to mind. There need not even be recognition (cognition) of the pupil's ignorance; there is only modulated behavior in response to the success or lack thereof of the naïve, with the result that the naïve learns more efficiently than otherwise. There is no intention of bringing about learning, and such behavior falls far short of customary concepts of teaching, undoubtedly present in ourselves.

Indeed, teaching in this differential behavior sense is found even in ants, when leaders lead followers to food (Franks and Richardson 2006). If we are going to accept such animal activities as (behavioral) teaching, then we need a modified account of (ideational) teaching, where teacher deliberately instructs disciple.

David Premack finds that humans are quite unique in this capacity to teach: "Teaching, which is strictly human, reverses the flow of information found in imitation. Unlike imitation, in which the novice observes the expert, the teacher observes the novice—and not only observes, but also judges and modifies" (Premack 2004:318). There is two-way positive and negative feedback, driven by approval and disapproval. In due course in human societies, the pupil likewise judges and modifies what the teacher teaches. In such recursive loops cumulative transmissible cultures can be endlessly generated and regenerated. Richard Bryne finds that chimpanzees may have glimmerings of other minds, but show little evidence of intentional teaching (Bryne 1995:141, 146, 154).

So humans are unique in their cultural capacities because the mind that enables culture can transcend genetics. Richard Lewontin puts it this way:

> Our DNA is a powerful influence on our anatomies and physiologies. In particular, it makes possible the complex brain that characterizes human beings. But having made that brain possible, the genes have made possible human nature, a social nature whose limitations and possible shapes we do not know except insofar as we know what human consciousness has already made possible. . . . History far transcends any narrow limitations that are claimed for either the power of the genes or the power of the environment to circumscribe us. . . . The genes, in making possible the development of human consciousness, have surrendered their power both to determine the individual and its environment. They have been replaced by an entirely new level of causation, that of social interaction with its own laws and its own nature. (Lewontin 1991:123)

Theodosius Dobzhansky, a pivotal figure in modern genetics, reflects:

> Human genes have accomplished what no other genes succeeded in doing. They formed the biological basis for a superorganic culture, which proved to be the most powerful method of adaptation to the environment ever developed by any species. . . . The development of culture shows regularities *sui generis*, not found in biological nature, just as biological phenomena are subject to biological laws that are different from, without being contrary to, the laws of inorganic nature. (Dobzhansky 1956:121–122)

Ian Tattersall concludes:

> With the arrival of behaviorally modern *Homo sapiens*, a totally unprecedented entity had appeared on Earth. . . . *Homo sapiens* is not simply an improved version of its ancestors—it's a new concept, qualitatively distinct from them. . . . It's more akin to an "emergent quality," whereby for chance reasons a new combination of features produces totally unexpected results. (Tattersall 1998:188–189)

By astronomical and evolutionary scales, the development of culture is many orders of magnitude more rapid, 5,000 years of human historical memories against 13 billion years of universal history, or 3.5 billion years of life on Earth. In recent centuries, the explosive speed of cultural innovation has increased, with an ever-enlarging knowledge base making possible technological innovation, owing in large part to the powers of science. In the recent decades, information accumulates and travels in culture at logarithmically increasing speeds. Today cultural development takes place digitized at megabytes per second over the Internet. We seem to have reached a turning point in the long accumulating story of cognition actualizing itself.

Symbolic Explosion: Human Language

The power of ideas in human life is as baffling as ever. The nature and origins of language are proving, according to some experts in the field "the hardest problem in science" (Christiansen and Kirby 2003; Hauser et al. 2002). Kuniyoshi L. Sakai finds: "The human left-frontal cortex is thus uniquely specialized in the syntactic processes of sentence comprehension, without any counterparts in other animals" (Sakai 2005:817). Spoken language requires the evolution of genes for producing speech as well as comprehending it, and such genes, differentiating humans from other primates, arose at a highly critical period in our evolution.

The best current estimates place the origin of linguistically capable humans in the range of 40,000 to 100,000 years ago (Appenzeller 1998; Holden 1998). The range itself reveals our ignorance of the origins of language. The FOXP2 gene, called a speech gene, arose less than 200,000 years ago and became the subject of strong selection, making language possible. Couple this with the genes enlarging our brain and the result is our mental incandescence.

Ideas pass from mind to mind, and for this hearing what is spoken is more important than sight—at least until the invention of writing. We already examined the co-option that made the evolution of hearing possible. Millennia later, written language (needing those eyes and their co-opted crystallins) has transformed cultures by making possible the transmission of thoughts nonorally, across centuries and peoples. Printing makes possible massive public communication, followed by radio, television, electronic communication, the Internet.

Humans co-opted earlier animal hearing and sounds to develop a discursive language in which words and texts have become powerful symbols of the world, of the logic of that world, and of our place in the world. Humans have a double-level orienting system: one in the genes, shared with animals in considerable part; another in the mental world of ideas, as this flowers forth from mind, for which there is really no illuminating biological analogue. There can be, so to speak, knowledge at a distance. When knowledge becomes "ideational," these "ideas" make it possible to conceptualize and care about what is not present to felt experience. Humans can produce arguments about ideals in the face of the real.

Cumulative transmissible cultures are made possible by the distinctive human capacities for language. Language "comes naturally" to us, in the sense that humans everywhere have it. The child picks up speech during normal development with marvelous rapidity; language acquisition is only more or less intentional. The child mind is innately prepared for such learning (Chomsky 1986). Human language, when it comes, is elevated remarkably above anything known in nonhuman nature. "The huge number of words that every child learns dwarfs the capabilities of the most sophisticated non-humans" (Hauser and Fitch 2003:159). Marc Hauser remarks that, cognitively, the difference between humans and chimps is greater than that between chimps and worms. "When we transform thoughts into speech, we do something that no other animal ever achieves" (Hauser and Bever 2008:1057). Our capacities for symbolization, abstraction, vocabulary development, teaching, literary expression, and argument are quite advanced; they do not come naturally as an inheritance from the other primates, whatever may otherwise be our genetic similarity with them. Though language comes naturally to humans, what is learned has been culturally transmitted, this or that specific language, and the content carried during childhood education is that of an acquired, nongenetic culture. On this language capacity the development, transmission, and criticism of culture depends.

In a major recent study of whether animals have language, the authors conclude: "It seems relatively clear, after nearly a century of intensive research on animal communication, that no species other than humans has a comparable capacity to recombine meaningful units into an unlimited variety of larger structures, each differing systemically in meaning." The primate communication "system apparently never takes on the open-ended generative properties of human language" (Hauser, Chomsky, and Fitch 2002:1576–1577).

Stephen R. Anderson, a linguist, concludes:

> When examined scientifically, human language is quite different in fundamental ways from the communication systems of other animals. . . . Using our native language, we can produce and understand sentences we have never encountered before, in ways that are appropriate to entirely novel circumstances. . . . Human languages have the property of including such a discrete infinity of distinct sentences because they are *hierarchical* and *recursive*. That is, the words of a sentence are not just strung out one after another, but are organized into

> phrases, which themselves can be constituents of larger phrases of the same type, and so on without any boundary. (Anderson 2004:2–8)

The result is "massive differences in expressive capacities between human language and the communicative systems of other animals" (Anderson 2004:11).

> No other primate functions communicatively in nature even at the level of protolanguage, and the vast gulf of discrete, recursive combinability must still be crossed to get from there to the language capacity inherent in every normal human. We seem to be alone on our side of that gulf, whatever the evolutionary path we may have taken to get there. (Anderson 2004:318)

After thirty years' study of communication in mountain gorillas, the researchers conclude:

> Gorilla close-calls [those made within the group] are very far from being language-like, they seem to be of the order of complexity of threat displays, as indeed do chimpanzee calls. That simplicity raises the question of why apes, popularly considered more intelligent than monkeys, have apparently a simpler mode of communication, in the sense that they apparently do not label the environment by association of specific calls with specific contexts. . . . We have no answer for the contrast. (Harcourt and Stewart 2001:257–258)

Cheney and Seyfarth (1990) found that vervet monkeys give different alarm signals for snakes, leopards, and eagles; other monkeys hear these different alarms and take cover appropriately to differing predators. So it seemed that the calling monkey intends to refer and communicate its knowledge to others. But the most recent evidence raises doubt whether the seeming "callers" intend to inform. Rather, these differing noises are just spontaneous response grunts in alarm, although other monkeys can learn from such differing sounds and respond appropriately to what predator is present. Such signals cannot "be considered as precursors for, or homologs of, human words."

"There is no evidence that calling is intentional in the sense of taking into account what other individuals believe or want" (Hauser, Chomsky, and Fitch 2002:1576). Even chickadees give out different calls when encountering predators of different sizes (Templeton et al. 2005). But it is a mistake to think this is either language or a precursor of language.

Chomsky concludes:

> There seems to be no substance to the view that human language is simply a more complex instance of something to be found elsewhere in the animal world. This poses a problem for the biologist, since, if true, it is an example of true "emergence"—the appearance of a qualitatively different phenomenon at a specific stage of complexity of organization. (Chomsky 1972:70)

Linguistic ideational uniqueness involves complex use of symbols. Ian Tattersall concludes: "We human beings are indeed mysterious animals. We are linked to the living world, but we are sharply distinguished by our cognitive powers, and much of our behavior is conditioned by abstract and symbolic concerns" (Tattersall 1998:3). Similarly, Richard Potts concludes:

> In discussing the evolution of human critical capacities, the overarching influence of symbolic activity (the means by which humans create meaning) is inescapable. Human cultural behavior involves not only the transmission of nongenetic information but also the coding of thoughts, sensations, and things, times, and places that are not visible. All the odd elaborations of human life, socially and individually, including the heights of imagination, the depths of depravity, moral abstraction, and a sense of God, depend on this *symbolic coding of the nonvisible.* (Potts 2004:263)

Neuroscience makes it difficult to imagine disembodied mind. Life is embedded in matter-energy, a singularity within it. Mind is embodied in biology, a singularity within it. Culture is always embodied in incarnate persons. Still, there is an aspect of ideational culture that can seem somewhat loose from material embodiment, perhaps analogous to the way in which many mathematicians are platonic. Ideas do transcend any particular instantiation in bodies, even in minds (like the idea of a triangle transcending any particular world triangle). Although an idea must be embodied in some material platform, there can be multiple platforms for the same idea (in mind, in book, on audio tape, in computer). The U.S. legal tradition of separation of church and state, for example, is "embodied," if you like, in Constitution and case law, but this is not exactly in anybody's head, or even in all the heads of the living generation. It is an ongoing ideational complex, stored in books and public records, as well as in particular minds. Similarly with the Golden Rule. And both dramatically affect human behavior.

Mind: Predictable or Surprising?

Might the mind's big bang have been anticipated? Again, biologists range across a spectrum. Despite finding other kinds of progress undeniable in the evolutionary record, Ernst Mayr reflects on the evolution of intelligence: "An evolutionist is impressed by the incredible improbability of intelligent life ever to have evolved" (1988:69). He elaborates this judgment:

> We know that the particular kind of life (system of macromolecules) that exists on Earth can produce intelligence. . . . We can now ask what was the probability of this system producing intelligence (remembering that the same system was able to produce eyes no less than 40 times). We have two large super-kingdoms of life on Earth, the prokaryote evolutionary lines each of which could lead theoretically to intelligence. In actual fact none of the thousands of lines among the prokaryotes came anywhere near it.
>
> There are 4 kingdoms among the eukaryotes, each again with thousands or ten thousands of evolutionary lineages. But in three of these kingdoms, the protists, fungi, and plants, no trace of intelligence evolved. This leaves the kingdom of Animalia to which we belong. It consists of about 25 major branches, the so-called phyla,

> indeed if we include extinct phyla, more than 30 of them. Again, only one of them developed real intelligence, the chordates. There are numerous Classes in the chordates, I would guess more than 50 of them, but only one of them (the mammals) developed real intelligence, as in Man. The mammals consist of 20-odd orders, only one of them, the primates, acquiring intelligence, and among the well over 100 species of primates only one, Man, has the kind of intelligence that would permit [the development of advanced culture]. Hence, in contrast to eyes, an evolution of intelligence is not probable. (Mayr, quoted in Barrow and Tipler 1986:132–133)

Mind of the human kind is unusual even on this unusual Earth. Biology is not inevitably anthropic.

If we move from Harvard to Cambridge, Simon Conway Morris takes an opposite view: "Something like ourselves is an evolutionary inevitability, and our existence also reaffirms our one-ness with the rest of Creation" (Conway Morris 2003:xv–xvi). "As all the principal properties that characterize humans are convergent, then sooner or later, and we still have a billion years of terrestrial viability in prospect, 'we' as a biological property will emerge" (96). He asks, and answers: whether,

> given time, evolution will inevitably lead not only to the emergence of such properties as intelligence, but also to other complexities, such as, say, agriculture and culture, that we tend to regard as the prerogative of the human? We may be unique, but paradoxically those properties that define our uniqueness can still be inherent in the evolutionary process. In other words, if we humans had not evolved then something more-or-less identical would have emerged sooner or later. (196)

Conway Morris continues: "'Hominization' is not as unique a process as many may think" (274). "Rerun the tape of life as often as you like, and the end result will be much the same. On Earth it happens to be humans" (282). Biology is anthropic.

Christian de Duve concludes that neural power, where it luckily arises, has such "decisive selective advantage" that there is high probability of its increase:

> The direction leading toward polyneuronal circuit formation is likely to be specially privileged in this respect, so great are the advantages linked with it. Let something like a neuron once emerge, and neuronal networks of increasing complexity are almost bound to arise. The drive toward larger brains and, therefore, toward more consciousness, intelligence, and communication ability dominates the animal limb of the tree of life on Earth. (de Duve 1995:297)

There is only one line that leads to persons, but in that line at least the steady growth of cranial capacity makes it difficult to think that intelligence is not being selected. One can first think that in humans enlarging brains are to be expected, since intelligence conveys obvious survival advantage. But then again, that is not so obvious, since all the other five million or so presently existing species survive well enough without advanced intelligence, as did all the other billions of species that have come and gone over the millennia. Reptiles were small-brained past and present, limited perhaps by being cold-blooded. Big brains take more energy, longer gestation, and more time rearing young. Mammal brains do grow but in fits and starts; generally mammals are four to five times brainer than reptiles.

In only one of these myriads of species has a brain capable of a transmissible culture developed, and in this one it has developed explosively. This does have survival advantage for humans, even if other species survive quite well without it. With only hunting and gathering technologies, humans became the most widespread mammal on Earth. The cumulation makes it possible for an individual to inherit the discoveries of thousands of others before him, discoveries that the individual could not make in a single lifetime.

But human cultures continue, characterized as well by radical innovations in cognition that eventually have little or nothing to do with survival. Grigori Perelman sought and found a proof to the Poincaré conjecture in mathematics, transforming irregular spaces into uniform ones (Mackenzie 2006). Ed Wilson cares for the conservation of his ants, where "splendor awaits in minute proportions" (1984:139). Humans do have to reproduce themselves in a next generation; but much, even most of what goes on in contemporary culture is not primarily focused on leaving more of one's genes in the next generation.

Are these outcomes in the third big bang latent in, inherent in the second big bang? If we look to science, the answer seems to be that scientists have reached no consensus. Conway Morris finds that humans or something much like us is inevitable; de Duve finds a natural selection of more brain in animal life, conveying survival advantage; but Mayr could not find any of this at all. John Maynard Smith too, we recall from the previous chapter, tracing the unique transitions that produced evolutionary escalation and led to humans, found "no reason to regard the unique transitions as the inevitable result of some general law"; to the contrary, these events might not have happened at all (Maynard Smith and Szathmáry 1995:3).

Let us return again to the question whether possibilities emerge en route. Molecules, trillions of them, spin around in this astronomically complex webwork and generate the unified, centrally focused experience of mind. This is a process for which we can as yet scarcely imagine a theory. Can we nevertheless say that the potential for intentionality was there at the big bang? Was self-reflexive mind, a thirteen-billion-years-later singularity, always lurking around from the startup? It was, earlier, quite a mental stretch to see the protons, electrons, and atoms waiting to string themselves together into DNA some nine or ten billion years later. Were they also waiting to form intentional mind? There is certainly nothing in contemporary physics that can address this problem; no "theory of everything" can explain artifacts of mind, such as a hammer or a cell phone, much less Perelman's proof.

Neither life nor mind is yet known elsewhere in the universe, but even if both were found elsewhere repeatedly, we would not know whether these repetitions were all front-loaded or whether new possibility spaces opened up in diverse places in the universe. Judgments in such cases might depend on how similar the extraterrestrial minds were, but of course we humans, even smart scientists, could only recognize minds with considerable similarity to our own. Life and mind elsewhere would not eliminate dimensions of unique particularity on Earth—elephants, humans, Israel, Einstein. The human mind seems to require an incredible opening up of new possibility space.

Consider this self-transcendence in the sciences—and now it is revealing to look beyond genetics and neuroscience, beyond the sciences where we study ourselves. Physics and astronomy are within our scientific cultures, and yet with these disciplines we transcend our cultures. With our instrumented intelligences and constructed theories, we now know of phenomena at structural levels from quarks to quasars. We measure distances from picometers to the extent of the visible universe in light-years, across forty orders of magnitude. We measure the strengths of the four major binding forces in nature (gravity, electromagnetism, the strong and weak nuclear forces), again across forty orders of magnitude. We measure time at ranges across thirty-four orders of magnitude, from attoseconds to the billions-of-years age of the universe. Nature gave us our mind-sponsoring brains; nature gave us our hands. Nature did not give us radiotelescopes with which to "see" pulsars, or relativity theory with which to compute time dilation. These come from human genius, cumulated in our transmissible cultures (though we do not forget that nature supplies these marvelous processes analyzed by radiotelemetry and relativity theory).

These extremes are beyond our embodied experience. No one experiences a light-year or a picosecond. But they are not beyond our comprehension entirely, else we could not use such concepts so effectively in science. The instrumentation is a construction (radiotelescopes and mathematics), a cultural invention, a "social construct," if you must. But precisely this construction enables us dramatically to extend our native ranges of perception. The construction dis-embodies us. It distances us from our embodiment. No one has an everyday "picture" of a quark or a pulsar. But we have good theory why nothing can be "seen" at such ranges in ordinary senses of "see," which requires light in the wavelength range of 400–700 nanometers, with the radiation from quarks and pulsars far outside that range. We can ask whether a molecule is too small to be colored, or whether an a electron, in its superposition states, is so radically different as to have no position, no "place" in the native range sense, only a probabilistic location.

That transcends startpoint location, enabling us to reach a standpoint location greater than ourselves. No animal, humans included, knows everything going on at all levels, quarks to cosmos ("the God's-eye view"). Some animals, sometimes humans, know little of what is going on at any level; they have only functional behaviors, genetically coded or behaviorally acquired, that work, more or less, for survival. They have, we might say, limited know-how and no know-that. But humans can sometimes enjoy an epistemic genius, transcending their own sector and take an overview (Earth seen from space, the planet's hydrologic cycles) or take in particulars outside their embodiment (sonar in bats, low-frequency elephant communication). We can, as we are doing here, consider these three big bangs, billion of years apart and spanning immense levels of complexity. This genius is fact of the matter. If science could find that this result is latent in the system, that would be a great marvel. If science can only find this outcome a serendipitous surprise in the system, that makes it no less a marvel. Either because of or despite their evolutionary origins, humans are a radically new kind of species on Earth.

Manfred Eigen concludes that there is transformation of the creative process:

> The process of creation is by no means at an end, although no-one can predict what is to come, even within intervals of time that are negligibly short in comparison with the phase of genetic evolution. . . . But evolutionary progress in the near future will hardly be on the genetic level. The activation of the human mind has greatly speeded up the roundabout of development. Almost everything that happens in the foreseeable future will proceed from mankind. . . . Man is still a relative newcomer to planet Earth, and the creation of humanity has only just begun. (Eigen 1992:49)

Life starts up, and, as already recognized, on many of its trajectories, it smarts up. That is as startling in the supersmart human head start as anywhere else in the universe. Blaisé Pascal's "thought" at the start of the Enlightenment is still true: "But, if a universe were to crush him, man would still be more noble than that which killed him, because he knows that he dies and the advantage which the universe has over him; the universe knows nothing of this. All our dignity consists, then, in thought" (Pascal [1670] 1958:# 347:97). As philosophers from ancient

Greece onward have claimed, humans are "the rational animals." Scientific research continues to confirm this ideational uniqueness. But deeper explanations require philosophy, metaphysics, theology. Ultimate explanations may require the dimension of "spirit."

Spirited Persons: The Ultimate Marvel

Humans find themselves uniquely emplaced on a unique planet—in their world cognitively and critically as no other species is. Our bodily incarnation embeds us in this biospheric community; we are Earthlings. Our mental genius enables us to rise to transcending overview. We humans are at once "spirited selves," enjoying our incarnation in flesh and blood, empowered for survival by our brain/minds, defending our personal selves, and yet transcending ourselves and our local concerns. *Homo sapiens* is the only part of the world free to orient itself with a view of the whole. We are not free from the worlds of either nature or culture, but free in those environments. That makes us, if you like, free spirits; it also makes us self-transcending spirits. That is the peculiar genius of the human "person" or "spirit." We alone can wonder where we are, who we are, what we ought to do.

A frequent way of reading the history of science displaces humans from central focus. Earth is a lonely planet, lost out there in the stars; humans are latecomers on Earth, arriving in the last few seconds of geological and astronomical time. We are cosmic dwarfs, trivial on the universe scale. Copernicus dealt a cosmological blow: humans do not live at the center of the universe. Darwin struck an evolutionary blow: humans are not divine but animals. Watson and Crick struck a molecular biology blow: humans are nothing but electronic molecules in motion on atomic scales. Freud struck a psychological blow, the most humiliating of all: we persons are not masters of our own minds.

But with a gestalt switch, one can read the same natural history to find cosmic genius in humans. This tension between decentering ourselves and humans at the center of interest is a framework or paradigm issue. In this perspective, humans live at the center of complexity. In astronomical nature and micronature, at both ends of the spectrum of size, nature lacks the complexity that it demonstrates at the mesolevels, found in at our native ranges on Earth. Perhaps we humans are cosmic dwarfs; perhaps we are molecular giants. But there is no denying our mid-scale complexity. We do not live at the range of the infinitely small, nor at that of the infinitely large, but we may well live at the range of the infinitely complex. The last big explosion in the universe is in your head.

Biologists anxious to displace any idea of direction or progress in evolutionary natural history may say that the tree of life is bushy, rather than treelike. Humans will be shown as one among other small buds on the bush, not some apical bud on top. Certainly we do wish to celebrate the diversity of life on Earth: the bushy explosion. But what this overlooks is the mental explosion in that bud on the bush, the radical way in which in one species the searching genes outdo themselves. Elaborating the genetic cybernetic possibilities, in generating human brains their genes crossed a threshold into a cognitive realm with spectacular new powers and freedoms. The combinatorial cybernetic explosion is recompounded. Terrence Deacon catches this uniqueness:

> Hundreds of millions of years of evolution have produced hundreds of thousands of species with brains, and tens of thousands with complex behavioral, perceptual, and learning abilities. Only one of these has ever wondered about its place in the world, because only one evolved the ability to do so. (Deacon 1997:21)

So we wonder where we are (cosmology, universe, Earth, creation), who we are (person, self, spirit, soul, made in the image of God), what we ought to do (ethics, justice, love, value choices). Let us begin with the latter, more immediately demanded in daily life, and move toward the former and more ultimate. The uniquely rational animal is equally the uniquely ethical animal. Ethics is distinctively a product of the human genius, a phenomenon in our social behavior. To be ethical is to reflect on considered principles of right and wrong and to act accordingly, in the face of temptation. This is a possibility in all and only human life, so that we expect and demand that persons behave morally and hold them so responsible. We approve and disapprove on principle and in practice. This is true even when, alas, some humans are tragically derelict or failing and we cannot presume to treat them as what they ought to have been, or perhaps once were, at least aspirationally. Such an emergence of ethics is as remarkable as any other event we know; in some form or other ethics is pervasively present in every human culture, whether honored in the observance or the breach.

We must attach loving to logic, if we are to understand the outcome of this third big bang. Reason is yoked with emotion, cognition with caring. The natural forces, thrusting up the myriad species, produced one that, so to speak, reached escape velocity, transcending the merely natural with cares super to anything previously natural. A complement of this eternal mystery is the possibility for better and worse caring, for noble and misplaced caring, for good and evil.

In this humans are unique; there is nowhere in animal behavior the capacity to be reflectively ethical. After a careful survey of behavior, Helmet Kummer concludes, "It seems at present that morality has no specific functional equivalents among our animal relatives" (Kummer 1980:45). Jerome Kagan puts it this way: "What is biologically special about our species is a constant attention to what is good and beautiful and a dislike of all that is bad and ugly. These biologically prepared biases rend the human experience incommensurable with that of any other species" (Kagan 1998:191). Peter Singer's *Ethics* has a section, "Common Themes in Primate Ethics," including a section on "Chimpanzee Justice," and he wants to "abandon the assumption that ethics is uniquely human" (Singer 1994:6). But many of the behaviors examined (helping behavior, dominance structures) are more pre-ethical than ethical; he has little or no sense of holding chimpanzees morally culpable or praiseworthy.

Frans de Waal finds precursors of morality, but concludes:

> Even if animals other than ourselves act in ways tantamount to moral behavior, their behavior does not necessarily rest on deliberations of the kind we engage in. It is hard to believe that animals weigh their own interests against the rights of others, that they develop a vision of the greater good of society, or that they feel lifelong guilt about something they should not have done. Members of some species may reach tacit consensus about what kind of behavior to tolerate

> or inhibit in their midst, but without language the principles behind such decisions cannot be conceptualized, let alone debated. (de Waal 1996:209)

As before with "culture" and with "teaching," finding "ethics" in nature is partly a matter of discovering previously unknown animal behavior, but mostly a matter of redefining and stretching what the word "ethics" means to cover behavioral adjustments in social groups.

Christopher Boehm finds that in some primate groups, not only is there a dominance hierarchy but also there are controls to keep such hierarchy working because this produces arrangements that the primates can live with, improving their overall success. Chimpanzees fight with each other over food and mates, but fighting is unpleasant, so the chimps will allow the dominant to break up such fights. If, however, the dominant becomes overly aggressive, the chimps will gang up on the dominant, who can control one but not several arrayed against him. The result is more "egalitarian behavior" (Boehm 1999). Perhaps such behaviors are the precursors out of which such maxims as "treat equals equally; treat unequals equitably" once emerged, but it must be equally clear that such chimps are orders of magnitude away from deliberate reflection on how to treat others fairly, respecting their rights, much less their dignity.

After her years of experience with chimpanzees, and although she finds pair bonding, grooming, and the pleasure of the company of others, Jane Goodall wrote:

> I cannot conceive of chimpanzees developing emotions, one for the other, comparable in any way to the tenderness, the protectiveness, tolerance, and spiritual exhilaration that are the hallmarks of human love in its truest and deepest sense. Chimpanzees usually show a lack of consideration for each other's feelings which in some ways may represent the deepest part of the gulf between them and us. (van Lawick-Goodall 1971:194)

Our primate relatives do not negotiate the presence of an existential self interacting interpersonally with other such agents in the process of thinking about and pursuing goals in the world. Higher animals realize that the behavior of other animals can be altered, and they do what they can to shape such behavior. So relationships evolve that set behavioral patterns in animal societies—dominance hierarchies, for example, or ostracism from a pack or troop. But it is not within the animal capacity to become a reflective agent interacting with a society of similar reflective agents, knowing that other actors (if normal), like oneself, are able to choose between options and are responsible for their behavior. Animals lack awareness that there are mental others whom they might hold responsible. Or to whom they might be held responsible. This precludes any critical sense of justice, of values that could and ought to be fairly shared because they are enjoyed by others who, like oneself, are existential subjects of their own lives. Such consideration is not a possibility in their private worlds, and neither is any morally binding social contract such as that in interhuman ethics. Yet all this, undeniably, has emerged within the human genius.

The demand that we be ethical pushes the further question of who we are. Persons set up a reflective gap between the real and ideal that orients action. Humans may desire, for instance, to preserve and enlarge family and tribe. We may come to care that democracy survives in the world, or that the wisdom of Shakespeare not be lost in the next generation, and work to fulfill such ideals. Persons may admire and try to be Good Samaritans. Cognitive science sometimes thinks of the mind as a kind of computer. But the embodied mind is not hardware, not software; it is (so to speak) wetware that must be kept wet, sometimes with tears, struggling to do right, surrounded by wrongs. A merely computational mind would be an incompetent judge of good and evil. "Temptation" is not found in computers. The human desires to be moral; however brokenly, ideal mixes with real.

This coupling of the ideal with broken embodiment amplifies in humans our earlier worries about struggle in the biological processes, whether and how far struggle is required for vital creativity. Perhaps from the first big bang, astronomical and chemical processes are fine-tuned, but the evolutionary epic cannot be. Now, at the level of personality, of spiritual creativity, fine-tuning seems even more irrelevant. One cannot fine-tune the adventures of incarnate minds navigating hyper-immense possibility space. Although Jesus would not have existed without the fine-tuned resonance states of carbon and oxygen, the life of Jesus was not fine-tuned, nor could it have been—whatever account a theologian might give of divine foreordination or his sense of destiny ("He set his face to go to Jerusalem," Luke 9.51). A parent—even a heavenly father—does not fine-tune the rearing of a son or daughter. Suffering love is never clockwork precision. If there is resonance, this is in sympathy and solidarity, spirit attuned to spirit, beset by hopes and fears in an ambiguous and challenging world.

Chemical reagents remain effective in human biochemistry, but spiritual agency, superimposed on this, is a radically new level of being. We find in persons an agent who must be oriented by a belief system, as, in the biological world, animals are not. That leaves us with the question of how to authorize such a belief system. In persons, the self-actualizing and self-organizing doubles back on itself with the qualitative emergence of what the Germans call "Geist," what existentialists call "Existenz." Matter can, the physicists say, be "excited" under radiation. The neural animal can, the biologists say, become "excited," emotional. Here, what is really "exciting" is that human intelligence is now "spirited," an ego with felt, psychological inwardness that cares about itself and its role in the world.

Persons have egos. They feel ashamed or proud; they have angst, self-respect, fear, and hope. They may get excited about a job well done, pass the buck for failures, have identity crises, or deceive themselves to avoid self-censure. Humans are capable of pride, avarice, flattery, adulation, courage, charity, forgiveness, prayer. They may resolve to dissent before an immoral social practice and pay the price of civil disobedience in the hope of reforming their society. They weep and say grace at meals. They may be overcome with anomie, or make a confession of faith. They may insult or praise each other. They tell jokes. Persons act in love, faith, or freedom, driven by guilt or seeking forgiveness—to use categories that theologians have thought fundamental.

Persons have unique careers that interweave to form storied narratives in cultural heritages. They have heroes or saviors who may die for the sins of the world, launch the Kingdom of God, or launch other passionate ideologies about the meanings of life and history. Persons may become disciples of these sages and saviors, and when they do they realize that to be a person includes a dimension of "spirit." Where there is reflective, sacrificial, suffering love, there is spirit. There is spirit where there is a sensing of the numinous, the sacred, the holy. There is spirit where there is awe, a sense of the sublime. There is spirit where, along with an explosion of knowledge, nature escalates as a wonderland. There is spirit when persons confront the limit questions, when persons get goose pimples looking into the night sky or at the Vishnu schist at the bottom of the Grand Canyon. Or pondering the three big bangs.

R. L. Stevenson pondered the "incredible properties" of dust stirring to give rise to this creature struggling for responsible caring:

> What a monstrous spectre is this man, the disease of the agglutinated dust, lifting alternate feet or lying drugged with slumber; killing, feeding, growing, bringing forth small copies of himself; grown upon with hair like grass, fitted with eyes that move and glitter in his face; a thing to set children screaming;—and yet looked at nearlier, known as his fellows know him, how surprising are his attributes! Poor soul, here for so little, cast among so many hardships, filled with desires so incommensurate and so inconsistent, savagely surrounded, savagely descended, irremediably condemned to prey upon his fellow lives: who should have blamed him had he been of a piece with his destiny and a being merely barbarous? And we look and behold him instead filled with imperfect virtues: infinitely childish, often admirably valiant, often touchingly kind; sitting down, amidst his momentary life, to debate of right and wrong and the attributes of the deity; rising up to do battle for an egg or die for an idea; singling out his friends and his mate with cordial affection; bringing forth in pain, rearing with longsuffering solicitude, his young. To touch the heart of his mystery, we find in him one thought, strange to the point of lunacy: the thought of duty; the thought of something owing to himself, to his neighbour, to his God: an ideal of decency, to which he would rise if it were possible; a limit of shame, below which, if it be possible, he will not stoop. (Stevenson 1903:291, 293–295)

The embodied story is the human legacy of waking up to good and evil (as in Genesis 1–2) or the dreams of hope for the future (as with visions of the Kingdom of God). This, as much as logic and love, may be the *differentia* of the human genius. The generation of such caring is as revealing as anything else we know about natural history. The fact of the matter is that evolution has generated ideals in caring.

Nor should we be surprised that this generating has been a long struggle. The evolutionary picture is of nature laboring in travail. The root idea in the English word "nature," going back to Latin and Greek origins, is that of "giving birth." Birthing is creative genesis, which certainly characterizes evolutionary nature. Birthing (as every mother knows) involves struggle. Earth slays her children, a seeming evil, but bears an annual crop in their stead. The "birthing" is nature's orderly self-assembling of new creatures amidst this perpetual perishing. Life is ever "conserved," as biologists might say; life is perpetually "redeemed," as theologians might say. Let us call it the "generation and regeneration of caring." Resulting from the second big bang, there is life, perpetually perishing, which can only be continued in explosion if there is struggling through, dying for the next generation. Resulting from the third big bang, there is mind, spirited mind acting in sacrificial love.

We contemporary humans, perhaps more than previous generations, have reached a critical turning point in the long-accumulating story of cognition actualizing itself. We are now coming around to oversee the world and to face the prospect of our own self-engineering, to the genesis of a higher-level ordering of the world in the midst of its threatening disorder. Increasingly we are like gods. But there is an information gap about good and evil. We need the wisdom of God, and that programs poorly on computers and is not found in physics, chemistry, or biology textbooks.

Presence with Presence

The singularities, if we may use a theological word, might also be "revealing" not simply about human spirit but about divine spirit, about "Presence." Science gives us three principal data points: matter-energy, life, and mind. The first is universal; the second is rare; the third is single and we are it. Surveying this trajectory from nature to spirit, we have repeatedly been wondering whether these three explosions are all somehow front-loaded into the system, or whether each is a one-off surprise. A frequent modern attitude is that before puzzlement one ought to be scientific about figuring things out. But we have found that no sciences settle what historical, much less what philosophical account to give of these three big bangs. Science is not well equipped to deal with singularities, one-off events; science prefers lawlike regularities.

Could the surprises have been anticipated? Each stage is necessary for the next, but no stage seems sufficient for the next. Each stage allows the next, but no stage logically implies the next. No scientific law, plus initial conditions, predicts each surprise. In some moods, the vast distances between the three, billions of years apart, suggest minimal connections. There is emergence, but is it driven or spontaneous? Even more provocatively, each stage launches escalating serendipity. Outrageous luck? Or are there "attractors"? Is there a subtending field, a deeper source?

We have ranged over a spectrum of options: random chance, probability, selective tendencies, necessity, design; and these permit some mixing. The formation of heavier elements in stars may have been inevitable, but the formation of a suitable planet may have been random chance. The launching of life may have been random chance, but, once launched, biodiversity was highly probable, biocomplexity less probable but likely. The formation of human mind may have been serendipitous, and after that cultural diversity may have been highly probable. Cultural diversity may be peculiar and local, or local on isolated islands but cumulative on continents and accelerated at crossroads between continents. The causal connections are likely themselves to be complex.

At such levels of complexity, we will often be in "over our heads," but one conclusion is inescapable: what is "in our heads" is as startling as anything else yet known in the universe. We will be left wondering how far what is going on "in our heads" is key, at cosmological and metaphysical levels, to what is going on "over our heads." Is mind key to the whole? Are we detecting Mind in, with, and under it all? We humans are spirited presence. Are we an icon of deeper Presence, Spirit suffusing the universe story?

Often those who had hoped to be scientific about answers will say that even if science does not give answers, we should still be naturalistic. Matter-energy, life, and mind are events in nature. With the evolution of each later stage, the tectonic potential of nature actualizes into something higher. Each of the emergent steps is "super" to the precedents, that is, supervenes on and surpasses the principles and processes earlier evident. Each transcends previous ontological levels. The category of the natural is elevated as it enlarges. Nature proves richer, more fertile, brooding, mysterious, than was recognized before. A spirited history, a history of spirit, supervenes on matter-energy. The generative power is lured toward spirit, evident in human spirits. And such nature is a supercharged nature, but still nature. There is no Supernature, but nature is super. Three big bangs document that. There is no God, but Nature should be spelled with a capital N, because Nature is sacred, the ground of our being. Looking ahead, this inexhaustible creative openendedness is greater than we now know, or can foreseeably know.

At this turn of thought, others may want still deeper explanations: a Transcendence in which this self-transcending nature is embedded, a Ground of all Being. Supercharged nature signals Transcendent Presence. The upper-level accounts cast their light back across what might in short-scope perspective have seemed complete naturalistic accounts. They cast shadows over them. The earlier events begin to figure as subplots within a larger story. Afterward, the naturalistic explanations do not look so compelling, as they earlier did. To believe in the supernatural is to believe that there are forces at work that transcend the physical, the biological, and the cultural. These spiritual forces sway the future because they have for millennia been breaking through and infusing what is going on. We may detect from our present vantage point intimations of a fourth dimension (Spirit) when three dimensions (matter, life, mind) are already incontestably evident and the fourth seems to be secretly and impressively also at work.

Almost anything can happen in a world in which what we see around us has actually managed to happen. The story is already incredible, progressively more so at every emergent level. Nature is indisputably there, and what are we to make of it? Both good induction and good historical explanation lead us to believe in surprises still to come and powers already at work greater than we know. For all the unifying theories of science, nature as a historical system has never yet proved simpler or less mysterious than we thought; the universe has always had more storied achievements taking place in it than we knew. To suspect the work of spiritual forces is not, in this view, to be naïve but rather to be realistic. This subtending Presence is eternal, equally co-present at the startup and en route.

Einstein concluded, famously, that "the eternal mystery of the world is its comprehensibility" (Einstein 1970:61). Going beyond Einstein, I am concluding that "the eternal mystery of the universe is its generating of comprehending, caring mind." In this sense, the astronomical, the evolutionary, the genetic, the neurological, and the psychological explosions all suggest that rational minds comprehending three big bangs may well believe that we inhabit a "spirited," a "spiritual" universe. We can wonder if there is a "Logos" in, with, and under the logic of such nature. Maybe we are not so lonely after all; our presence is embraced by another Presence.

REFERENCES

Alroy, John, et al. 2008. "Phanerozoic Trends in the Global Diversity of Marine Invertebrates." *Science* 321 (4 July): 97–100.

Alumets, J., R. Hakanson, F. Sundler, and J. Thorell. 1979. "Neuronal Localisation of Immunoreactive Enkephalin and f3-endorphin in the Earthworm." *Nature* 279 (28 June): 805–806.

Anderson, P. W. 1972. "More Is Different." *Science* 177 (4 August): 393–396.

Anderson, Stephen R. 2004. *Doctor Dolittle's Delusion: Animals and the Uniqueness of Human Language*. New Haven: Yale University Press, 2005.

Appenzeller, Tim. 1998. "Art: Evolution or Revolution." *Science* 282:1451–1454.

Arendt, Detlev, Alexandru S. Denes, Gáspár Jékely, and Kristin Tessmar-Raible. 2008. "The Evolution of Nervous System Centralization." *Philosophical Transactions of the Royal Society B* 363:1523–1528.

Ayala, Francisco J. 1974. "The Concept of Biological Progress." In Francisco Jose Ayala and Theodosius Dobzhansky, eds., *Studies in the Philosophy of Biology*. New York: Macmillan.

Bains, William. 2004. "Many Chemistries Could Be Used to Build Living Systems." *Astrobiology* 4:137–167.

Bak, Per. 1997. *How Nature Works: The Science of Self-Organized Criticality*. New York: Springer-Verlag.

Balashov, Yuri V. 1991. "Resource Letter AP-1: The Anthropic Principle." *American Journal of Physics* 59:1069–1076.

Barr, Stephen M. 2003. *Modern Physics and Ancient Faith*. Notre Dame, IN: University of Notre Dame Press.

Barrow, John D. 2002. *The Constants of Nature*. New York: Pantheon.

—. 2004. "Cosmology and Immutability." In John D. Barrow, Paul C. W. Davies, and Charles L. Harper, Jr., eds., *Science and Ultimate Reality: Quantum Theory, Cosmology, and Complexity*, 402–425. Cambridge: Cambridge University Press.

—. 2007. *New Theories of Everything: The Quest for Ultimate Explanation*. New York: Oxford University Press.

Barrow, John D., Simon Conway Morris, Stephen J. Freeland, and Charles L. Harper Jr. 2008. *Fitness of the Cosmos for Life: Biochemistry and Fine-Tuning*. Cambridge: Cambridge University Press.

Barrow, John D. and Joseph Silk. 1980. "The Structure of the Early Universe." *Scientific American* 242, no. 4 (April).

Barrow, John D. and Frank J. Tipler. 1986. *The Anthropic Cosmological Principle*. New York: Oxford University Press.

Bear, Mark F., Barry W. Connors, and Michael A. Paradiso. 2001. *Neuroscience: Exploring the Brain*, 2nd ed. Baltimore: Lippincott Williams and Wilkins.

Benton, M. J. 1995. "Diversification and Extinction in the History of Life." *Science* 268 (7 April):52–58.

Boehm, Christopher. 1999. *Hierarchy in the Forest: The Evolution of Egalitarian Behavior*. Cambridge, MA: Harvard University Press.

Boesch, Christophe. 1996. "The Emergence of Culture among Wild Chimpanzees." *Proceedings of the British Academy* 88:251–268.

Bonner, John Tyler. 1988. *The Evolution of Complexity by Means of Natural Selection*. Princeton: Princeton University Press.

Boyd, Robert and Peter J. Richerson. 1985. *Culture and the Evolutionary Process*. Chicago: University of Chicago Press.

Brading, Katherine and Elena Castellani. 2003. *Symmetries in Physics: Philosophical Reflections*. Cambridge: Cambridge University Press.

Bray, Dennis. 2009. *Wetware: A Computer in Every Living Cell*. New Haven: Yale University Press.

Brockman, John. 1995. *The Third Culture: Beyond the Scientific Revolution*. New York: Simon and Schuster.

Bryne, Richard. 1995. *The Thinking Ape: Evolutionary Origins of Intelligence*. New York: Oxford University Press.

Burger, William C. 2003. *Perfect Planet, Clever Species: How Unique Are We?* Amherst, NY: Prometheus.

Burtt, E. A. 1996. *The Metaphysical Foundations of Modern Physical Science* (1952). Atlantic Highlands, NJ: Humanities Press.

Campbell, Donald T. 1974. "'Downward Causation' in Hierarchically Organised Biological Systems." In Francisco Jose Ayala and Theodosius Dobzhansky, eds., *Studies in the Philosophy of Biology: Reduction and Related Problems*, 179–186. Berkeley: University of California Press.

Caro, T. M. and M. D. Hauser. 1992. "Is There Teaching in Nonhuman Animals?" *Quarterly Review of Biology* 67, no. 2:151–174.

Carr, Bernard, ed. 2007. *Universe or Multiverse?* Cambridge: Cambridge University Press.

Carr, Bernard J. and Martin J. Rees. 1979. "The Anthropic Principle and the Structure of the Physical World," *Nature* 278 (12 April): 605–612.

Chaplin, Martin F. 2001. "Water: Its Importance to Life." *Biocemistry and Molecular Biology Education* 29:54–59.

Cheney, Dorothy L. and Robert M. Seyfarth. 1990. *How Monkeys See the World*. Chicago: University of Chicago Press.

Chimpanzee Sequencing and Analysis Consortium, 2005. "Initial Sequence of the Chimpanzee Genome and Comparison with the Human Genome." *Nature* 437:69–87.

Chomsky, Noam. 1972. *Language and Mind*. New York: Harcourt Brace Jovanovich.

——. 1986. *Knowledge of Language: Its Nature, Origin, and Use*. New York: Praeger Scientific.

Christiansen, Morten H. and Simon Kirby. 2003. "Language Evolution: The Hardest Problem in Science?" In Christiansen and Kirby, eds., *Language Evolution*, 1–15. New York: Oxford University Press.

Clayton, Donald D. 1983. *Principles of Stellar Evolution and Nucleosynthesis*. Chicago: University of Chicago Press.

Conway Morris, Simon. 2003. *Life's Solution: Inevitable Humans in a Lonely Universe*. Cambridge: Cambridge University Press

Conway Morris, Simon and Stephen Jay Gould. 1998. "Showdown on the Burgess Shale." *Natural History* 107, no. 10:48–55.

Cowan, W. Maxwell. 1979. "The Development of the Brain." *Scientific American* 241, no. 3 (September): 112–133.

Crick, Frances. 1981. *Life Itself*. New York: Simon and Schuster.

Darwin, Charles. 1968. *The Origin of Species* (1859). Baltimore: Pelican Books.

Day, William. 1984. *Genesis on Planet Earth: The Search for Life's Beginning*. 2nd ed. New Haven: Yale University Press.

Davies, Paul C. W. 1982. *The Accidental Universe*. New York: Cambridge University Press.

Davies, Paul. 2007. *Cosmic Jackpot: Why Our Universe Is Just Right for Life*. Boston: Houghton Mifflin.

Dawkins, Richard. 1995. *River Out of Eden: A Darwinian View of Life*. New York: Basic Books, HarperCollins.

Deacon, Terrence W. 1997. *The Symbolic Species: The Co-evolution of Language and the Brain*. New York: Norton.

de Duve, Christian. 1995. *Vital Dust: The Origin and Evolution of Life on Earth*. New York: Basic Books.

——. 1996. "The Birth of Complex Cells." *Scientific American* 274, no. 4:50–57.

——. 2002. *Life Evolving: Molecules, Mind, and Meaning*. New York: Oxford University Press.

——. 2005. *Singularities: Landmarks on the Pathways of Life*. Cambridge: Cambridge University Press.

Dennett, Daniel C. 1987. *The Intentional Stance*. Cambridge, MA: MIT Press.

Denton, Michael. 1998. *Nature's Destiny: How the Laws of Biology Reveal Purpose in the Universe*. New York: Free Press.

de Waal, Frans. 1996. *Good Natured: The Origins of Right and Wrong in Humans and Other Animals*. Cambridge, MA: Harvard University Press.

——. 1999. "Cultural Primatology Comes of Age," *Nature* 399(17 June):635–636.

Dickson, Barry J. 2002. "Molecular Mechanisms of Axon Guidance." *Science* 298:1959–1964.

Dobzhansky, Theodosius. 1956. *The Biological Basis of Human Freedom*. New York: Columbia University Press.

——. 1967. *The Biology of Ultimate Concern*. New York: New American Library.

Dorus, Steve et al. 2004. "Accelerated Evolution of Nervous System Genes in the Origin of *Homo sapiens*." *Cell* 119:1027–1040.

Draganski, Bogdan et al. 2004. "Changes in Grey Matter Induced by Training." *Nature* 427 (22 January): 311–312.

Dyall, Sabrina D., Mark T. Brown, and Patricia J. Johnson. 2004. "Ancient Invasions: From Endosymbionts to Organelles." *Science* 304:253–257.

Eigen, Manfred. 1971. "Selforganization of Matter and the Evolution of Biological Macromolecules." *Die Naturwissenschaften* 58:465–523.

Eigen, Manfred with Ruthild Winkler-Oswatitsch. 1992. *Steps Towards Life: A Perspective on Evolution*. New York: Oxford University Press.

Einstein, Albert. 1970. *Out of My Later Years*. Rev. reprint ed. Westport, CT: Greenwood Press.

Eisemann, C. H., W. K. Jorgensen, D. J. Meritt, M. J. Rice, B. W. Cribb, P. D. Webb, and M. P. Zalucki. 1984. "Do Insects Feel Pain?—A Biological View." *Experientia* 40:164–167.

Elbert, Thomas et al. 1995. "Increased Cortical Representation of the Fingers of the Left Hand in String Players." *Science* 270 (13 October): 305–307.

Ellis, George F. R. 2005. "Physics, Complexity and Causality." *Nature* 435 (9 June): 743.

Enard, Wolfgang et al. 2002. "Intra- and Interspecific Variation in Primate Gene Expression Patterns." *Science* 296 (12 April): 340–343.

Erwin, Douglas H. 1993. *The Great Paleozoic Crisis: Life and Death in the Permian*. New York: Columbia University Press.

Finney, John L. 2004. "Water: What's So Special About It?" *Philosophical Transactions of the Royal Society: Biological Sciences* 359:1145–1165.

Flannagan, Owen. 1992. *Consciousness Reconsidered*. Cambridge, MA: MIT Press.

Fodor, J. 1992. "The Big Idea: Can There Be a Science of Mind?" *The Times Literary Supplement*, July 3, 5–7.

Franks, Nigel R. and Tom Richardson. 2006. "Teaching in Tandem-running Ants." *Nature* 439 (12 January): 153.

Fuster, Joaquín M. 2003. *Cortex and Mind: Unifying Cognition*. New York: Oxford University Press.

Gabora, Liane. 2006. "Self-Other Organization: Why Early Life Did Not Evolve Through Natural Selection." *Journal of Theoretical Biology* 241:443–450.

Galef, Bennett G., Jr. 1992. "The Question of Animal Culture." *Human Nature* 3, no. 2: 157–178.

García-Ruiz, Juan Manuel, Emilio Melero-García, and Stephen T. Hyde. 2009. "Morphogenesis of Self-Assembled Nanocrystalline Materials of Barium Carbonate and Silica." *Science* 323 (16 January): 362–365.

Gardner, Martin. 2003. *Are Universes Thicker Than Blackberries?* New York: Norton.

Gazzaniga, Michael S. 2008. *Human: The Science Behind What Makes Us Unique*. New York: Ecco.

Ghysen, Alain. 2003. "The Origin and Evolution of the Nervous System." *International Journal of Developmemtal Biology* 47:555–562.

Gianaro, Catherine. 2005. "Human Cognitive Abilities Resulted from Intense Evolutionary Selection, Says Lahn." *The University of Chicago Chronicle* 24, no. 7 (January 6): 1, 5.

Gould, Stephen Jay. 1983. "Extemporaneous Comments on Evolutionary Hope and Realities." In Charles L. Hamrum, ed., *Darwin's Legacy, Nobel Conference XVIII*, 95–103. San Francisco: Harper and Row.

——. 1989. *Wonderful Life: The Burgess Shale and the Nature of History*. New York: Norton.

Gould, Stephen Jay and Niles Eldredge. 1977. "Punctuated Equilibria: The Tempo and Mode of Evolution Reconsidered." *Paleobiology* 3:115–151.

Graesser, Michael L., Stephen D. H. Hsu, Alejandro Jenkins, and Mark B. Wise. 2004. "Anthropic Distribution for Cosmological Constant and Primordial Density Pertubations." *Physics Letters B* 600:15–21.

Gross, David J. 1996. "The Role of Symmetry in Fundamental Physics." *Proceedings of the National Academy of Sciences, USA* 93:14256–14259.

Guth, Alan. 1997. *The Inflationary Universe: The Quest for a New Theory of Cosmic Origins*. Reading, MA: Helix Books.

Harcourt, Alexander H. and Kelly J. Stewart. 2001. "Vocal Relationships of Wild Mountain Gorillas." In Martha M. Robbins, Pascale Sicotte, and Kelly

J. Stewart, eds., *Mountain Gorillas: Three Decades of Research at Karisoke*, 241–262. Cambridge: Cambridge University Press.

Hauser, Marc D. and Thomas Bever. "A Biolinguistic Agenda." *Science* 322 (14 November 2008): 1057–1059.

Hauser, Marc D., Noam Chomsky, and W. Tecumseh Fitch. 2002. "The Faculty of Language: What Is It, Who Has It, and How Did It Evolve?" *Science* 298 (22 November): 1569–1579.

Hauser, Marc D. and W. Tecumseh Fitch. 2003. "What Are the Uniquely Human Components of the Language Faculty?" In Morten H. Christiansen and Simon Kirby, eds., *Language Evolution*, 158–181. New York: Oxford University Press.

Hawking, Stephen. 1998. *A Brief History of Time*. Updated and expanded 10th anniversary ed. New York: Bantam Books.

——. 1996. "Quantum Cosmology." In Stephen Hawking and Roger Penrose, *The Nature of Space and Time*, 75–103. Princeton: Princeton University Press.

Helitzer, Florence. 1973. "The Princeton Galaxy." *Intellectual Digest* 3, no. 10 (July 1973): 25–32.

Hirth, F. and H. Reichart. 1998. "Basic Nervous System Types: One or Many?" In Jon H. Kaas, ed., *Evolution of Nervous Systems*, vol. 1, 55–72. San Diego: Academic Press.

Holden, Constance. 1998. "No Last Word on Language Origins." *Science* 282: 1455–1458.

Holderness, Mike. 2001. "Think of a Number." *New Scientist* 170 (16 June): 45.

Horgan, John. 1996. *The End of Science: Facing the Limits of Knowledge in the Twilight of the Scientific Age*. Reading, MA: Addison-Wesley.

Hoyle, Fred. 1981. "The Universe: Past and Present Reflections." *Engineering and Science* 45, no. 2 (November): 8–12.

Jacob, François. 1977. "Evolution and Tinkering." *Science* 196:1161–1166.

Jaspers, Karl. 1953. *The Origin and Goal of History*. New Haven: Yale University Press.

Jennings, H. S. 1906. *Behavior of the Lower Organisms*. New York: Columbia University Press.

Kagan, Jerome. 1998. *Three Seductive Ideas*. Cambridge: Harvard University Press.

Kauffman, Stuart A. 2007. "Beyond Reductionism: Reinventing the Sacred." *Zygon: Journal of Religion and Science* 42:903–914.

Kazazian, Jr., Haig H. 2004. "Mobile Elements: Drivers of Genome Evolution." *Science* 303:1626–1632.

Kimura, Motoo. 1961. "Natural Selection as the Process of Accumulating Genetic Information in Adaptive Evolution." *Genetical Research* 2:127–140.

Kirschner, Marc W. and John C. Gerhart. 1998. "Evolvability." *Proceedings of the National Academy of Sciences, USA* 95:8420–8427.

——. 2005. *The Plausibility of Life*. New Haven: Yale University Press.

Knoll, Andrew H. 1986. "Patterns of Change in Plant Communities Through Geological Time." In Jared Diamond and Ted J. Case, eds., *Community Ecology*, 126–141. New York: Harper and Row.

——. 2003. *Life on a Young Planet*. Princeton: Princeton University Press.

Kummer, Helmut, 1980. "Analogs of Morality Among Nonhuman Primates." In Gunter Stent, ed., *Morality as a Biological Phenomenon*, 31–47. Berkeley: University of California Press.

Kunz, Werner and Matthias Kellermeier. 2009. "Beyond Biomineralization." *Science* 323 (16 January): 344–345.

Kvenvolden, Keith, et al. 1970. "Evidence for Extraterrestrial Amino-Acids and Hydrocarbons in the Murchison Meteorite." *Nature* 228:923–926.

Laughlin, R. B. and David Pines. 2000. "The Theory of Everything." *PNAS (Proceedings of the National Academy of Sciences)* 97 (January 4): 28–31.

LeDoux, Joseph. 2002. *Synaptic Self: How Our Brains Become Who We Are*. New York: Viking.

Leslie, John. 1989. *Universes*. London: Routledge.

Levy, Yaakov and José N. Onuchic. 2006. "Water Mediation in Protein Folding and Molecular Recognition." *Annual Review of Biophysics and Biomolecular Structure* 35:389–415.

Lewontin, R. C. 1991. *Biology as Ideology: The Doctrine of DNA*. New York: HarperCollins.

Linde, A. D. 1990. *Inflation and Quantum Cosmology*. San Diego: Academic Press.

Lloyd, Seth. 2006. *Programming the Universe*. New York: Knopf.

Lorenz, Edward N. 1968. "Climatic Determinism." *Meteorological Monographs* 8, no. 30: 1–3.

Lovell, Bernard. 1975. "Whence?" *New York Times Magazine*, November 16, pp. 27, 72–90, 95. This is a shortened form of "In the Centre of Immensities," Lovell's Presidential Address to the British Association for the Advancement of Science, August 27, 1975.

Maciá, E., M. V. Hernández, and J. Oró. 1997. "Primary Sources of Phosphorus and Phosphates in Chemical Evolution." *Origins of Life and Evolution of the Biosphere* 27:459–480.

Mackenzie, Dana. 2006. "The Poincaré Conjecture Proved." *Science* 314 (22 December): 1848–1849.

Maguire, Eleanor A. et al. 2000. "Navigation-Related Structural Change in the Hippocampi of Taxi Drivers." *Proceedings of the National Academy of Sciences of the United States of America* 97, no. 8:4398–4403.

Mather, Jennifer A. and Roland C. Anderson. 1993. "Personalities of Octopuses (*Octopus rubescens*)." *Journal of Cognitive Psychology* 107:336–340.

Matsuzawa, Tetsuro. 2007. "Comparative Cognitive Development." *Developmental Science* 10:97–103.

Maynard Smith, John. 1995. "Life at the Edge of Chaos?" *New York Review of Books* 52, no. 4 (March 2): 28–30.

—. 2000. "The Concept of Information in Biology." *Philosophy of Science* 67:177–194.

—. 2000. "Reply to Commentaries." *Philosophy of Science* 67:214–218.

Maynard Smith, John and Eörs Szathmáry. 1995. *The Major Transitions in Evolution*. New York: W. H. Freeman.

Mayr, Ernst. 1982. *The Growth of Biological Thought: Diversity, Evolution and Inheritance*. Cambridge, MA: Harvard University Press.

—. 1988. *Toward a New Philosophy of Biology*. Cambridge, MA: Harvard University Press.

Mead, Margaret. 1989 "Preface." In Ruth Benedict, *Patterns of Culture* (1959), xi–xiv. Boston: Houghton Mifflin.

Mechelli, Andrea et al. 2004. "Structural Plasticity in the Bilingual Brain," *Nature* 431 (14 October): 757.

Merzenich, Michael. 2001. "The Power of Mutable Maps." In Mark F. Bear, Barry W. Connors, and Michael A. Paradiso, *Neuroscience: Exploring the Brain*, 2nd ed., box essay, 418. Baltimore: Lippincott Williams and Wilkins.

Monod, Jacques. 1972. *Chance and Necessity*. New York: Random House.

Nave, C. R. 2002. "Big Bang Expansion and the Fundamental Forces." *HyperPhysics*, http:\\hyperphysics.phy-astr.gsu.edu/hbase/astro/unify.html (accessed August 24, 2008).

Niklas, Karl J. 1986. "Large-Scale Changes in Animal and Plant Terrestrial Communities." In D. M. Raup and D. Jablonski, eds., *Patterns and Processes in the History of Life*, 383–405. New York: Springer-Verlag.

Niklas, Karl J., Bruce H. Tiffney, and Andrew H. Knoll. 1985. "Patterns in Vascular Land Plant Diversification: An Analysis at the Species Level." In James W. Valentine, ed., *Phanerozoic Diversity Patterns: Profiles in Macroevolution*, 97–128. Princeton: Princeton University Press.

Noble, Denis. 2006. *The Music of Life*. New York: Oxford University Press.

Oberhummer, H., A. Csótó, and H. Schlattl. 2000. "Stellar Production Rates of Carbon and Its Abundance in the Universe." *Science* 289 (7 July): 88–90.

Pace, Norman R.. 2001. "The Universal Nature of Biochemistry." *Proceedings of the National Academy of Sciences, U.S.A.* 98:805–808.

Padian, K. and W. A. Clemens. 1985. "Terrestrial Vertebrate Diversity: Episodes and Insights." In James W. Valentine, ed., *Phanerozoic Diversity Patterns: Profiles in Macroevolution*, 41–96. Princeton: Princeton University Press.

Pantev, Christo et al. 1998. "Increased Auditory Cortical Representation in Musicians." *Nature* 392 (23 April): 811–814.

Pascal, Blaise. 1958. *Pascal's Pensées* (1670). New York: E. P. Dutton.

Pennisi, Elizabeth. 1999. "Are Our Primate Cousins 'Conscious'?" *Science* 284:2073–2076.

——. 2006. "Brain Evolution on the Far Side." *Science* 314 (13 October): 244–245.

Penrose, Roger. 2005. *The Road to Reality: A Complete Guide to the Laws of the Universe*. New York: Knopf.

Perry, D. A., M. P. Amaranthus, J. G. Borchers, S. L. Borchers, and R. E. Brainerd. 1989. "Bootstrapping in Ecosystems." *BioScience* 39:230–237.

Perutz, M. F. 1983. "Species Adaptation in a Protein Molecule." *Molecular Biology and Evolution* 1:1–28.

Pielou, E. C. 1975. *Ecological Diversity*. New York: Wiley.

Pilbeam, D. 1972. *The Ascent of Man: An Introduction to Human Evolution*. New York: Macmillan.

Polkinghorne, John and Nicholas Beale. 2009. *Questions of Truth*. Louisville, KY: Westminster/John Knox Press.

Popper, Karl R. 1990. *A World of Propensities*. Bristol, England: Thoemmes.

Potts, Richard. 2004. "Sociality and the Concept of Culture in Human Origins." In Robert W. Sussman and Audrey R. Chapman, eds., *The Origins and Nature of Sociality*, 249–269. New York: Aldine de Gruyter.

Povinelli, Daniel J., Jesse M. Bering, and Steve Giambrone. 2000. "Toward a Science of Other Minds: Escaping the Argument from Analogy." *Cognitive Science* 24:509–541.

Povinelli, Daniel J. and Jennifer Vonk. 2003. "Chimpanzee Minds: Suspiciously Human?" *Trends in Cognitive Sciences* 7, no. 4: 157–160.

Premack, David. 1985. "'Gavagai!' or the Future History of the Animal Language Controversy." *Cognition* 19:207–296.

——. 2004. "Is Language the Key to Human Intelligence." *Science* 303:318–320.

Prigogine, Ilya and Stengers, Isabelle. 1984. *Order out of Chaos: Man's New Dialogue with Nature*. New York: Bantam Books.

Rees, Martin. 2000. *Just Six Numbers: The Deep Forces That Shape the Universe*. New York: Basic Books.

——. 2001. *Our Cosmic Habitat*. Princeton, NJ: Princeton University Press.

Rendell, Luke and Hal Whitehead. 2001. "Culture in Whales and Dolphins." *Behavioral and Brain Sciences* 24:309–382.

Richerson, Peter J. and Robert Boyd. 2005. *Not by Genes Alone: How Culture Transformed Human Evolution*. Chicago: University of Chicago Press.

Ringe, Dagmar and Gregory A. Petsko. 2008. "How Enzymes Work." *Science* 320 (13 June): 1428–1429.

Ruse, Michael. 1986. *Taking Darwin Seriously*. Oxford: Basil Blackwell.

——. 1996. *Monad to Man: The Concept of Progress in Evolutionary History*. Cambridge, MA: Harvard University Press.

Sakai, Kuniyoshi L. 2005. "Language Acquisition and Brain Development." *Science* 310 (4 November): 815–819.

Scott, Alwyn, 1995. *Stairway to the Mind: The Controversial New Science of Consciousness*. New York: Copernicus; Springer-Verlag.

——. 2002. *Neuroscience: A Mathematical Primer*. New York: Springer-Verlag.

Shapiro, James A. 2002. "Genome System Architecture and Natural Genetic Engineering." In Laura F. Landweber and Erik Winfree, eds., *Evolution as Computation*, 1–14. New York: SpringerVerlag.

——. 2005. "A 21st Century View of Evolution: Genome System Architecture, Repetitive DNA, and Natural Genetic Engineering." *Gene* 345:91–100.

Signor, Philip W., 1990. "The Geologic History of Life." *Annual Review of Ecology and Systematics* 21:509–539.

Simpson, George Gaylord. 1967. *The Meaning of Evolution*, rev. ed. New Haven: Yale University Press

Singer, Peter. 1994. *Ethics*. New York: Oxford University Press.

Skinner, H. Catherine W. 2002. "In Praise of Phosphates, or Why Vertebrates Chose Apatite to Mineralize Their Skeletal Elements." In W. G. Ernst, ed., *Frontiers in Geochemistry: Organic, Solution, and Ore Deposit Geochemistry*, vol. 2, 41–49. Columbia, MD: Bellwether Publishing and Geological Society of America.

Smolin, Lee. 1997. *The Life of the Cosmos*. Oxford: Oxford University Press.

Stanley, Steven M. 2007. "An Analysis of the History of Marine Animal Diversity." *Paleobiology* 33, no. 4 (supplement): 1–55.

Sterelny, Kim and Paul E. Griffiths. 1999. *Sex and Death: An Introduction to Philosophy of Biology*. Chicago: University of Chicago Press.

Stevenson, Robert Louis. 1903. "Pulvis et Umbra." In *Across the Plains*, 289–301. New York: Charles Scribner's Sons.

Stoeger, William R. 2007. "Reductionism and Emergence: Implications for the Interaction of Theology with the Natural Sciences." In Nancey Murphy and William R. Stoeger, eds., *Evolution and Emergence: Systems, Organisms, Persons*, 229–247. New York: Oxford University Press.

Tattersall, Ian. 1998. *Becoming Human: Evolution and Human Uniqueness*. New York: Harcourt Brace.

Taylor, Paul D. and Gilbert P. Larwood, eds. 1990. *Major Evolutionary Radiations*. Oxford: Clarendon Press.

Tegmark, Max. 1998. "Is 'the Theory of Everything' Merely the Ultimate Ensemble Theory?" *Annals of Physics* 270:1–51.

——. 2003. "Parallel Universes." *Scientific American* 288, no. 5 (May): 30–41.

Templeton, Christopher N., Erick Greene, and Kate Davis. 2005. "Allometry of Alarm Calls: Black-Capped Chickadees Encode Information About Predator Size." *Science* 308 (24 June): 1934–1937.

Thornton, Alex and Katherine McAuliffe. 2006. "Teaching in Wild Meerkats." *Science* 313 (14 July): 227–229.

Tomasello, Michael. 1999. *The Cultural Origins of Human Cognition*. Cambridge, MA: Harvard University Press.

Tomasello, Michael, Ann Cale Kruger, and Hilary Horn Ratner. 1993. "Cultural Learning." *Behavioral and Brain Sciences* 16, no. 3: 495–552.

Tye, Michael. 1997. "The Problem of Simple Minds: Is There Anything It Is Like to Be a Honey-Bee?" *Philosophical Studies* 88:289–317.

Tylor, E. B. 1903. *Primitive Cultures.* 4th ed., 2 vols. London: John Murray.

Valentine, James W. 1969. "Patterns of Taxonomic and Ecological Structure of the Shelf Benthos During Phanerozoic Time." *Palaeontology* 12:684–709.

——. 1973. *Evolutionary Paleoecology of the Marine Biosphere*. Englewood Cliffs, NJ: Prentice-Hall.

——. 2004. *On the Origin of Phyla*. Chicago: University of Chicago Press.

van Lawick-Goodall, Jane. 1971. *In the Shadow of Man*. Boston: Houghton Mifflin.

Venter, J. Craig et al. 2001. "The Sequence of the Human Genome." *Science* 291 (16 February): 1304–1351.

Vermeij, Geerat J. 1987. *Evolution and Escalation: An Ecological History of Life*. Princeton: Princeton University Press.

Vickers-Rich, Patricia and Patricia Komarower, eds. 2007. *The Rise and Fall of the Ediacaran Biota*. Bath, UK: The Geological Society.

Vilenkin, Alexander. 2006. "The Vacuum Energy Crisis." *Science* 312:1148–1149.

Vogel, Gretchen. 1999. "Chimps in the Wild Show Stirrings of Culture." *Science* 284 (24 June): 2070–2073.

von Baeyer, Hans Christian. 2003. *Information: The New Language of Science*. London: Weidenfeld and Nicolson.

Ward, Peter D. and Donald Brownlee. 2000. *Rare Earth: Why Complex Life Is Uncommon in the Universe.* New York: Copernicus; Springer-Verlag.

Weinberg, Steven. 1988. *The First Three Minutes*. New York: Basic Books.

——. 2002. "Is the Universe a Computer?" *New York Review of Books* 49, no. 16 (October 24): 43–47.

Westheimer, F. H. 1987. "Why Nature Chose Phosphates." *Science* 235 (6 March): 1173–1178.

Wheeler, John A. 1975. "The Universe as Home for Man." In Owen Gingerich, ed. *The Nature of Scientific Discovery,* 261–296. Washington: Smithsonian Institution Press.

——. 1994. *At Home in the Universe*. Woodbury, NY: AIP Press, American Institute of Physics.

——. 1999. "Information, Physics, Quantum: The Search for Links." In Anthony J. G. Hey, ed., *Feynman and Computation: Exploring the Limits of Computers,* 309–336. Reading, MA: Perseus Books.

Whiten, Andrew. 2005. "The Second Inheritance System of Chimpanzees and Humans." *Nature* 437 (1 September): 52–55.

Whitesides, George M. 2008. "Foreword: The Improbability of Life." In John D. Barrow, Simon Conway Morris, Stephen J. Freeland, and Charles L.

Harper, Jr., eds., *Fitness of the Cosmos for Life: Biochemistry and Fine-Tuning*, xi–xix. Cambridge: Cambrige University Press.

Whittaker, R. H. 1972. "Evolution and Measurement of Species Diversity." *Taxon* 21:213–251.

Wiener, Norbert. 1948. *Cybernetics*. New York: Wiley.

Wigner, Eugene P. 1960. "The Unreasonable Effectiveness of Mathematics in the Natural Sciences." *Communications on Pure and Applied Mathematics* 13 (1960): 1–14.

Wilczek, Frank. 1999. "Getting Its from Bits." *Nature* 397:303–306.

Williams, Robert J. P. 1953. "Metal Ions in Biological Systems." *Biological Reviews* 28:381–415.

—. 2000. "The Inorganic Chemistry of Life." In Nina Hall, ed., *The New Chemistry*, 259–299. Cambridge: Cambridge University Press.

Wilson, Edward O. 1975. *Sociobiology: The New Synthesis*. Cambridge, MA: Harvard University Press.

—. 1978. *On Human Nature*. Cambridge, MA: Harvard University Press.

—. 1984. *Biophilia*. Cambridge, MA: Harvard University Press.

—. 1992. *The Diversity of Life*. Cambridge, MA: Harvard University Press.

Wistow, Graeme. 1993. "Lens Crystallins: Gene Recruitment and Evolutionary Dynamism." *Trends in Biochemical Sciences* 18:301–306.

Wolfram, Stephen. 2002. *A New Kind of Science*. Champaign, IL: Wolfram Media.

Wrangham, Richard W., W. C. McGrew, Frans B. M. De Waal, and Paul G. Heltne, eds. 1994. *Chimpanzee Cultures*. Cambridge, MA: Harvard University Press.

Yokey, Hubert P. 2005. *Information Theory, Evolution, and the Origin of Life*. Cambridge: Cambridge University Press.

Zimmer, Carl. 2003. "How the Mind Reads Other Minds." *Science* 300 (16 May): 1079–1080.

INDEX